PORTABLE

WORKSHOP™

*Basic Wood Projects with Portable Power Tools*

# Decorative Accessories

Minnetonka, Minnesota, USA

# Credits


Cowles Creative Publishing, Inc.
Formerly Cy DeCosse Incorporated
5900 Green Oak Drive
Minnetonka, Minnesota 55343
1-800-328-3895

Printed in U.S.A.

*President:* Iain Macfarlane
*Executive V.P.:* William B. Jones
*Group Director, Book Development:* Zoe Graul

*Executive Editor:* Paul Currie
*Editorial Director:* Bryan Trandem
*Managing Editor:* Kristen Olson
*Associate Creative Director:* Tim Himsel
*Lead Project Designer:* Richard Steven
*Writers and Editors:* Craig Gelderman, Karl Larson, Richard Steven, Andrew Sweet
*Editor & Lead Technical Artist:* Jon Simpson
*Lead Art Director:* David Schelitzche
*Keyliners:* Eileen Bovard, Dan Liesen, Bob Rose, Gina Seeling
*Technical Production Editor:* Greg Pluth
*Project Designer & Technical Checker:* Rob Johnstone
*Project Designer:* John Gingerich
*Technical Art Draftsman:* John T. Drigot
*Vice President of Photography & Production:* Jim Bindas
*Copy Editors:* Janice Cauley, Hazel Jensen
*Builders:* Troy Johnson, Rob Johnstone
*Production Staff:* Curt Ellering, Laura Hokkanen, Paulette Johnson, Greg Wallace, Kay Wethern
*Studio Services Manager:* Marcia Chambers
*Photo Services Coordinator:* Cheryl Neisen
*Lead Photographer:* Rebecca Schmitt
*Prop Stylists:* Diane Heath, Coralie Sathre
*Production Manager:* Kim Gerber

*Printed on American paper by:*
Inland Press 00 99 98 97 / 5 4 3 2 1

*Created by:* The Editors of Cowles Creative Publishing, Inc., in cooperation with Black & Decker.


Library of Congress
Cataloging-in-Publication Data

Decorative accessories.
p. cm. -- (Portable workshop)
At head of title: Black & Decker.
ISBN 0-86573-643-x.

1. Furniture making--Amateurs' manuals.
2. House furnishings--Amateurs' manuals.
I. Cowles Creative Publishing.
II. Black & Decker Corporation (Towson, Md.)
III. Series.
TT195.D43 1997
684.1'04--dc21 97-24170

# Contents

# Introduction

How we decorate our living spaces speaks volumes about who we are. Spending five minutes in someone's home can reveal more about their values, interests and tastes than a day-long conversation. Because home decorating is a deeply personal act, we would all like to choose decorative accessories that accurately reflect our unique individuality.

But in our era of standardized mass production, where can you find unique decorative accessories without spending an astronomical amount? The answer is simple: create them yourself.

*Decorative Accessories* provides nearly two dozen creative projects that can help turn your house into a unique home. Each project reflects a thoughtful design and includes clear, easy-to-understand instructions. Although some projects are relatively elaborate, every design in this book can be built using only basic hand and portable power tools. We provide the instruction, you provide the creativity and energy. Together, we can help create the home you want without spending a lot of money.

For easy construction and your convenience, we used only common products sold in most building centers and corner hardware stores. Each of the 21 projects in this book consists of a complete cutting list, a lumber-shopping list, a detailed construction drawing, full-color photographs of major steps, and directions that guide you through every step of the project. One note, however. If you have never used any of the tools before, it's a good idea to first practice using them on scraps of wood. And, of course, follow the safety instructions.

The Black & Decker® Portable Workshop™ book series gives weekend do-it-yourselfers the power to build beautiful wood projects without spending a lot of money. Ask your local bookseller for more information on other volumes in this innovative new series.

NOTICE TO READERS

*This book provides useful instructions, but we cannot anticipate all of your working conditions or the characteristics of your materials and tools. For safety, you should use caution, care, and good judgment when following the procedures described in this book. Consider your own skill level and the instructions and safety precautions associated with the various tools and materials shown. Neither the publisher nor Black & Decker® can assume responsibility for any damage to property, injury to persons, or losses incurred as a result of misuse of the information provided.*

## Organizing Your Worksite

Portable power tools and hand tools offer a level of convenience that is a great advantage over stationary power tools. But using them safely and conveniently requires some basic housekeeping. Whether you are working in a garage, a basement or outdoors, it is important that you establish a flat, dry holding area where you can store tools. Set aside a piece of plywood on sawhorses, or dedicate an area of your workbench for tool storage, and be sure to return tools to that area once you are finished with them. It is also important that all waste, including lumber scraps and sawdust, be disposed of in a timely fashion. Check with your local waste disposal department before throwing away any large scraps of building materials or any finishing-material containers.

***Safety Tips***

*•Always wear eye and hearing protection when operating power tools and performing any other dangerous activities.*

*•Choose a well-ventilated work area when cutting or shaping wood and when using finishing products.*

## Tools & Materials

At the start of each project, you will find a set of symbols (see below) that show which power tools you will need. For some projects, a router table or power miter box, available at rental centers, will also be helpful. You will also need a set of basic hand tools: a hammer, screwdrivers, tape measure, a level, a combination square, C-clamps, and pipe or bar clamps. You will also find a shopping list of all the construction materials you will need. Miscellaneous materials and hardware are listed with the cutting list that accompanies the construction drawing. When buying lumber, note that the "nominal" size of the lumber is usually larger than the "actual size." For example, a 2 × 4 is actually 1½ × 3½".

## Power Tools You Will Use

**Circular saw** *to make straight cuts. For long cuts and rip-cuts, use a straightedge guide. Install a carbide-tipped combination blade for most projects.*

**Drills:** *use a cordless drill for drilling pilot holes and counterbores, and to drive screws; use an electric drill for sanding and grinding tasks.*

**Jig saw** *for making contoured cuts and internal cuts. Use a combination wood blade for most projects where you will cut pine, cedar or plywood.*

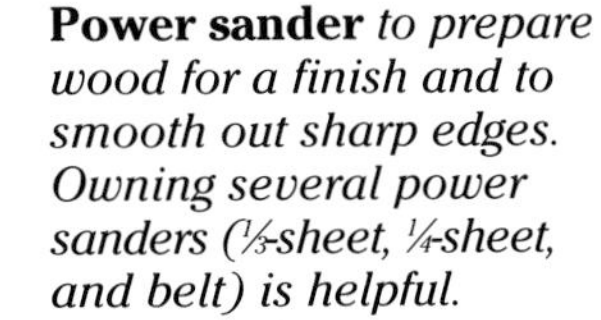

**Power sander** *to prepare wood for a finish and to smooth out sharp edges. Owning several power sanders (⅓-sheet, ¼-sheet, and belt) is helpful.*

**Belt sander** *for resurfacing rough wood. Can also be used as a stationary sander when mounted on its side on a flat worksurface.*

**Router** *to cut decorative edges and roundovers in wood. As you gain more experience, use routers for cutting grooves (like dadoes) to form joints.*

### *Guide to Building Materials Used in This Book*

***•Sheet goods:***

*OAK PLYWOOD: Oak-veneered plywood commonly sold in ¾" and ¼" thicknesses. Fairly expensive.*
*BIRCH PLYWOOD: Has smooth surface excellent for painting or staining; few voids in the edges. Moderately expensive.*
*MDF (MEDIUM-DENSITY FIBERBOARD): Plywood with a pressed-wood core that is well suited for shaping. Moderately inexpensive.*
*PINE PANELS: Pine boards glued together, cut and sanded. Varying thicknesses, usually ⅝" or ¾".*
*HARDBOARD: Basically, dense cardboard with a hard surface. Sold in ⅛" and ¼" thickness. Good for making back panels and other nonstructural parts. Very inexpensive.*

***•Dimension lumber:***

*BIRCH: A dense, heavy wood, highly resistant to wear. Relatively inexpensive.*
*WALNUT: A dark-colored hardwood with a coarse texture that takes finishes very well. Often used for furnishings. Moderate.*
*POPLAR: A soft hardwood suitable for painting. Inexpensive.*
*RED OAK: A common hardwood that stains well and is very durable. Relatively inexpensive.*
*CEDAR: Excellent outdoor wood with rich, warm color. Moderate.*

### *Guide to Fasteners & Adhesives Used in This Book*

***•Fasteners & hardware:***

*WOOD SCREWS: Brass or steel; most projects use screws with a #6 or #8 shank. Can be driven with a power driver.*
*DECK SCREWS: Galvanized for weather resistance. Widely spaced threads for good gripping power in soft lumber.*
*NAILS & BRADS: Finish nails can be set below the wood surface: common (box) nails have wide, flat heads; brads or wire nails are very small, thin fasteners with small heads.*
*MISCELLANEOUS HARDWARE: Door and drawer pulls; a variety of hinges; staples; shelf supports; hanger plates; roller catches; birch shaker pegs.*

***•Adhesives:***

*WOOD GLUE: Yellow glue is suitable for all interior projects.*
*MOISTURE-RESISTANT WOOD GLUE: Any exterior wood glue, such as plastic resin glue.*

***•Miscellaneous materials:***

*Wood plugs (for filling counterbores); dowels; Plexiglas®; glass; shelf nosing; acrylic; retaining clips; decorative trim moldings; veneer tape.*

## Finishing Your Project

Before applying finishing materials, fill nail holes and blemishes with wood putty or filler. Also, fill all voids in the edges of any exposed plywood with wood putty. Insert wood plugs into counterbore holes, then sand until the plug is level with the wood. Sand wood surfaces with medium sandpaper (100- or 120-grit), then finish-sand with fine sandpaper (150- or 180-grit). Wipe off residue, and apply the finish of your choice. Apply two or three thin coats of a hard, protective topcoat, like polyurethane, over painted or stained wood.

# Candle-sticks

PROJECT POWER TOOLS

*Place these candle-sticks on a table, windowsill or mantel, and watch them light up your life.*

CONSTRUCTION MATERIALS

| Quantity | Lumber |
|---|---|
| 1 | ½" × 1 × 2' Baltic birch plywood |
| 1 | 1 × 6" × 2' birch |

We considered a wide variety of shapes and styles before deciding on this handsome design for our candlesticks. More than 11" tall, these candlesticks are both elegant and inexpensive.

Appearances can be deceiving. These candlesticks look like the work of a highly skilled woodworker, but in reality they are quite simple to build. The shaped pedestals follow a simple pattern that can be transferred to your workpiece and then cut with a jig saw. The smooth edges on the bases and tops are made with a router table and a roundover bit. A handy jig simplifies the task of attaching the pedestals to the bases. And "candle nails" hold the candles in place.

Made from inexpensive Baltic birch plywood, these candlesticks can be painted with any decorative finish you choose.

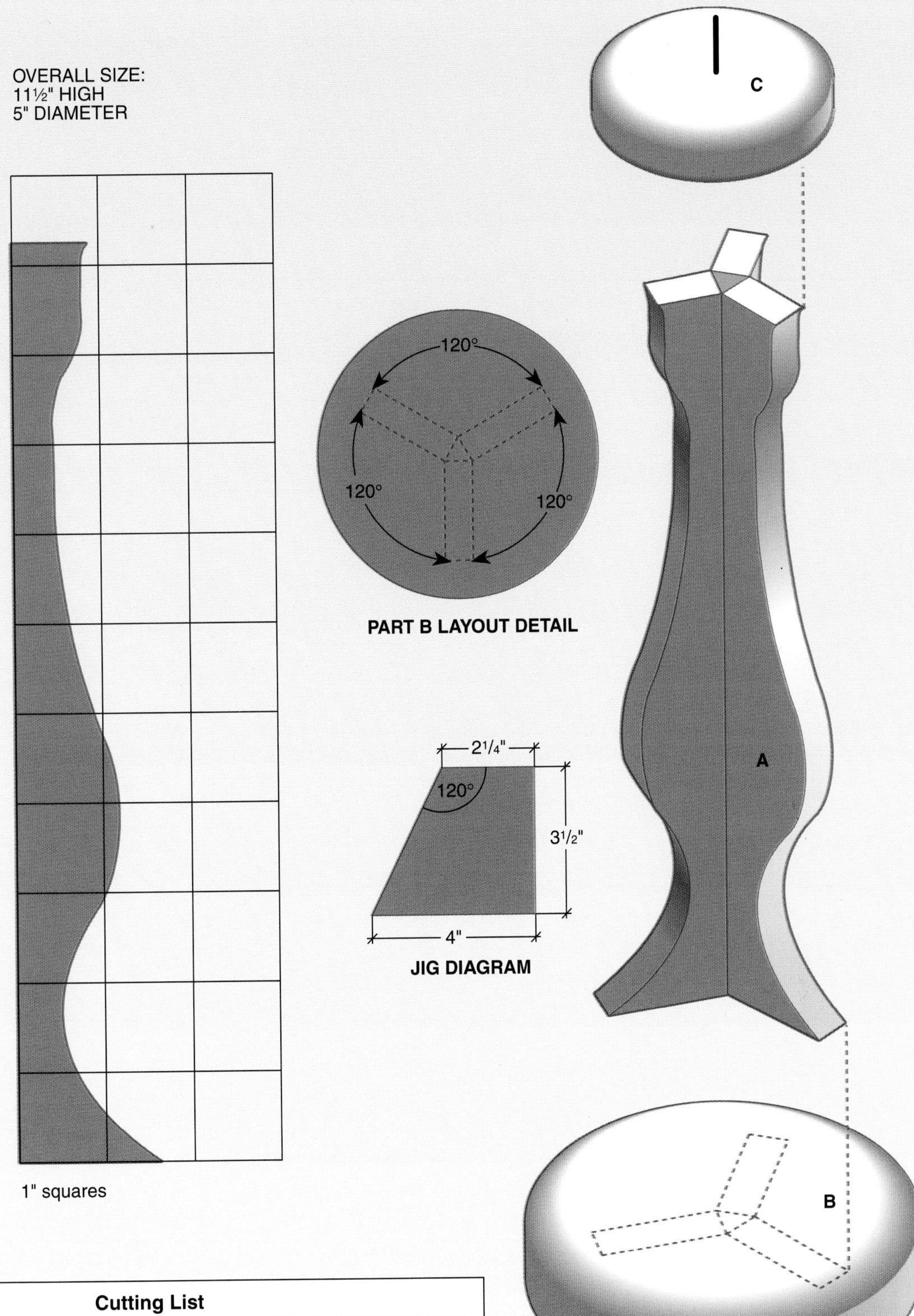


**Cutting List**

| Key | Part | Dimension | Pcs. | Material |
|---|---|---|---|---|
| **A** | Pedestal | ½ × 1¾ × 10¼" | 6 | Baltic birch plywood |
| **B** | Base | ¾ × 5" | 2 | Birch |
| **C** | Top | ½ × 3½" | 2 | Baltic birch plywood |

**Materials:** Wood glue, wood screws (#8 × 1½"), 1" brads, 1¼" roofing nails, finishing materials.

**Note:** Measurements reflect the actual thickness of dimensional lumber.

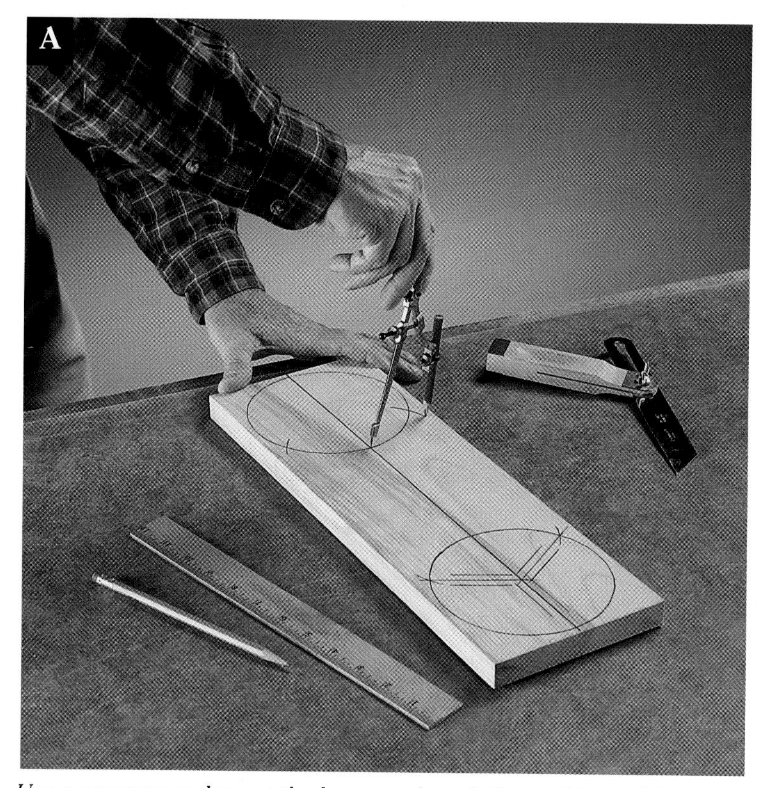


*Use a compass to lay out the bases and mark the positions of the pedestals on the bases. Use a sliding T-bevel to make sure the angles are uniform.*

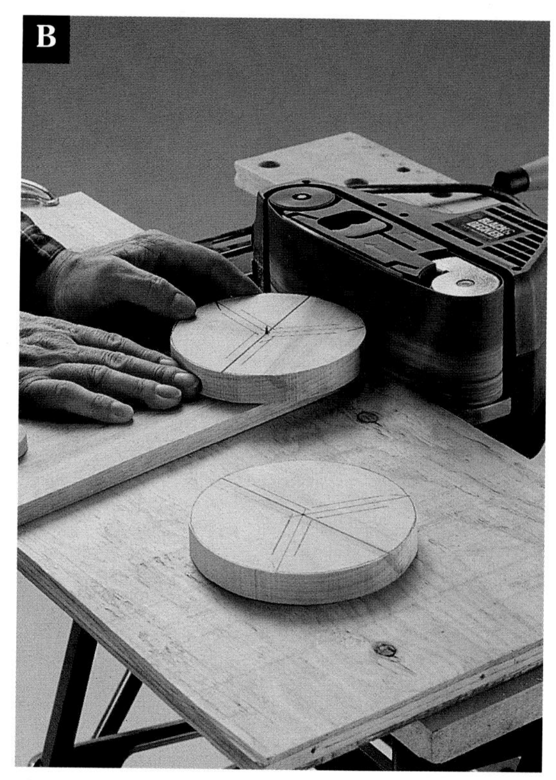


*Sand the edges of the bases with a belt sander and nail jig.*

## Directions: Candlesticks

MAKE THE BASES. Laying out the bases and the pedestal locations requires some basic geometry and the use of a drafting compass. Practice on paper until you're confident you understand the technique.

Start by drawing a center reference line down the length of a piece of 1 × 6" birch. Use a compass to draw two 2½"-radius circles, with centerpoints positioned on the reference line.

With the compass still set at 2½", place the tip at one of the points where the reference line crosses the circle, and scribe two arcs intersecting the circle. Draw lines connecting each of these intersection points to the centerpoint of the circle. These lines, along with the original reference line, form the three centerlines for the pedestal pieces. To mark the outline of the pedestal pieces, draw parallel lines ¼" on each side of the three centerlines. Repeat this process for the other circle **(photo A).** Use a jig saw to cut out the bases (B).

Build a sanding jig from a scrap piece of ¾" wood. Drill a pilot hole no more than 1½" from the end of the scrap board. Also drill a pilot hole through the centerpoint of the base. Drive a 1¾" nail through the pilot hole in the jig and into the base. Clamp a belt sander perpendicular to your worksurface and sand the edge of the base by rotating it on the nail **(photo B).** Repeat for the other base.

MAKE THE TOPS AND SHAPE THE PIECES. Use a compass set to 1¾" radius to draw two circles on ½" Baltic birch, and cut out the tops (C) with a jig saw. To sand the edges of each top, drill pilot holes through the centerpoint and use the nail jig and your belt sander.

Using a router table and a ¼" roundover bit, shape both edges of the tops and the top edges of the bases **(photo C).**

BUILD AND ATTACH THE PEDESTALS. Build the placement jig (see *Diagram*) to use as a guide when positioning the pedestal pieces.

Cut the pedestal blanks (A) to size. Transfer the pattern (see *Diagram*) to the blanks and cut with a jig saw. Clamp the pieces together, attach a drum sander to your drill and gang-sand the edges of the pedestals.

**TIP**

*When transferring a grid pattern, you have two options. You can enlarge the pattern on a copier and trace it onto a piece of cardboard to form a tracing template. Or, you can grid your stock with 1" squares and draw the pattern by hand directly onto the workpieces.*

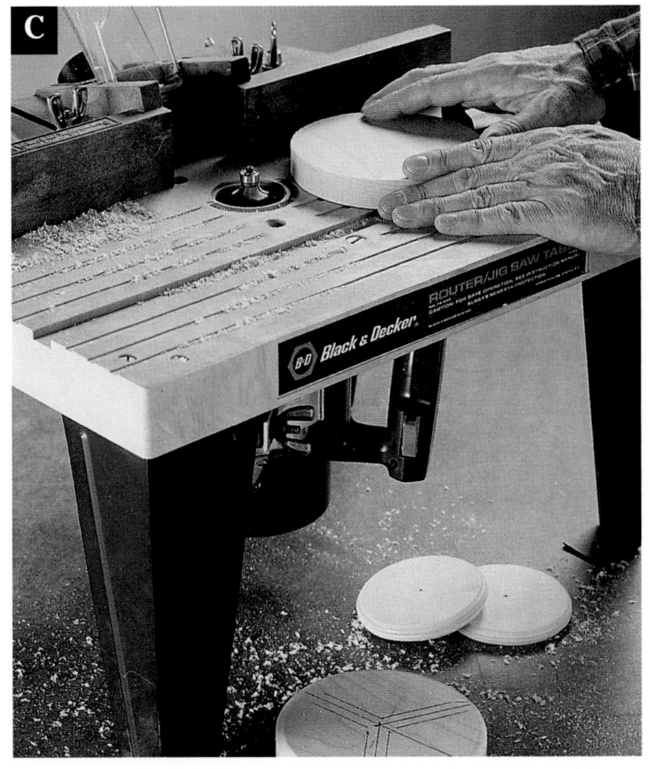


*Shape the edges of the tops and bases with a router table and a ¼" piloted roundover bit.*

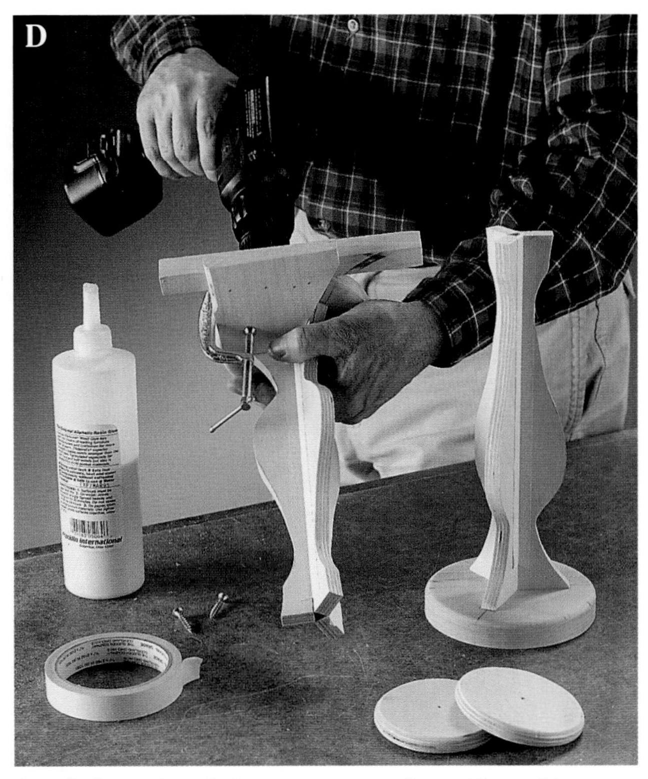


*Attach the pedestal pieces to one another with masking tape when joining them to the base.*

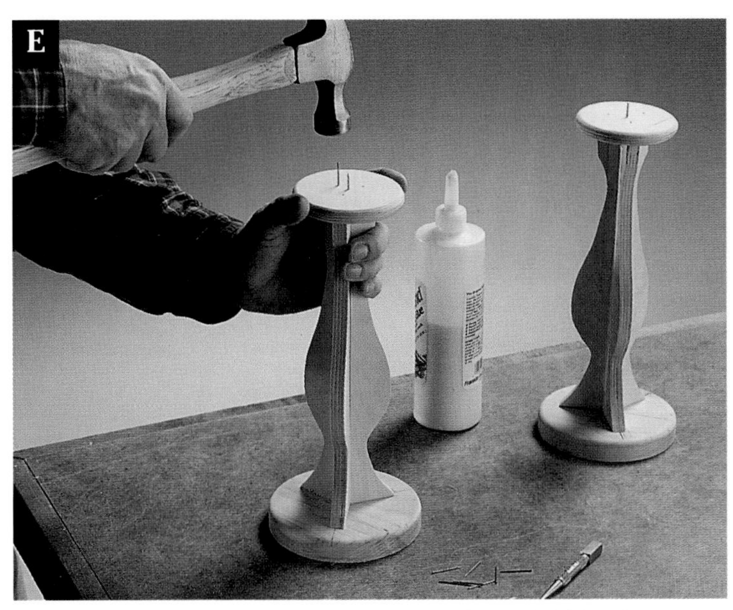


*Attach the top to the pedestals with wood glue and 1" brads.*

Using the pedestal outlines as a guide, drill pilot holes through the base, then countersink the holes from the bottom side. Position two pedestals on your worksurface so the back edges are against each other and the top and bottom edges are flush. Join the pedestals with a strip of masking tape. Fold the left pedestal over the right and align the third pedestal along the back edge of the left pedestal and join with tape.

Align the 120° angle of the placement jig along the outlines of two adjacent pedestal pieces, and clamp it to the base. Attach the pedestals with glue and screws driven up through the bottom face of the base, holding the pedestals firmly against the jig **(photo D).** After attaching two pedestals, reposition the jig and attach the third pedestal.

ATTACH THE TOPS. Drive a "candle nail" (we used a 1¼" roofing nail) up through the pilot hole in the center of each top. Center the tops on the pedestals, drill pilot holes and attach them with glue and 1" brads **(photo E).** Set the nail heads.

APPLY FINISHING TOUCHES. Scrape off any excess glue, and fill nail holes with putty. Finish-sand the candlesticks, and paint as desired (we used a "burled mahogany" faux paint kit).

# Exterior Light Post

*Your yard, deck, or patio will sparkle in the evening with this rustic outdoor light post.*

PROJECT POWER TOOLS

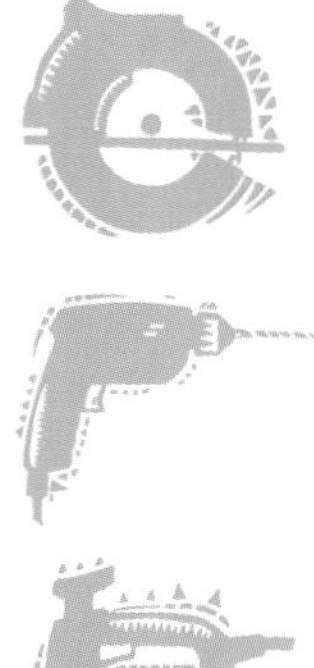

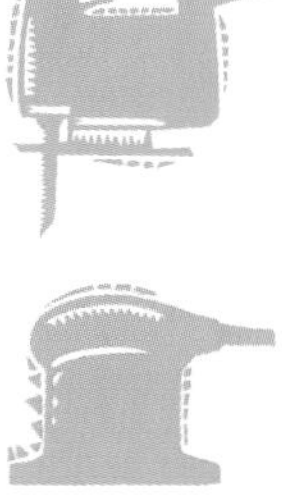

CONSTRUCTION MATERIALS

| Quantity | Lumber |
|---|---|
| 2 | 1 × 8" × 6' cedar |
| 1 | 1 × 10" × 2' cedar |
| 1 | 1/16" × 2 × 2' acrylic |

Lighting is an important but often overlooked element in landscape design. Our sturdy light post brings increased safety, security and attractiveness to any outdoor setting.

Made of durable, easy-to-machine cedar boards, this light post can be positioned to highlight plantings, illuminate walkways or to let you check on outdoor noises at the flip of a switch.

The top of our light post is held firmly in place with invisible roller catches, which make changing bulbs a snap.

Imagine yourself grilling on the patio, welcoming evening guests or just gazing at the summer stars—all in the soft glow of these unobtrusive sentinels.

Whether you decide to make one light post or several, this easy-to-build project will provide years of enjoyment for both family and guests.

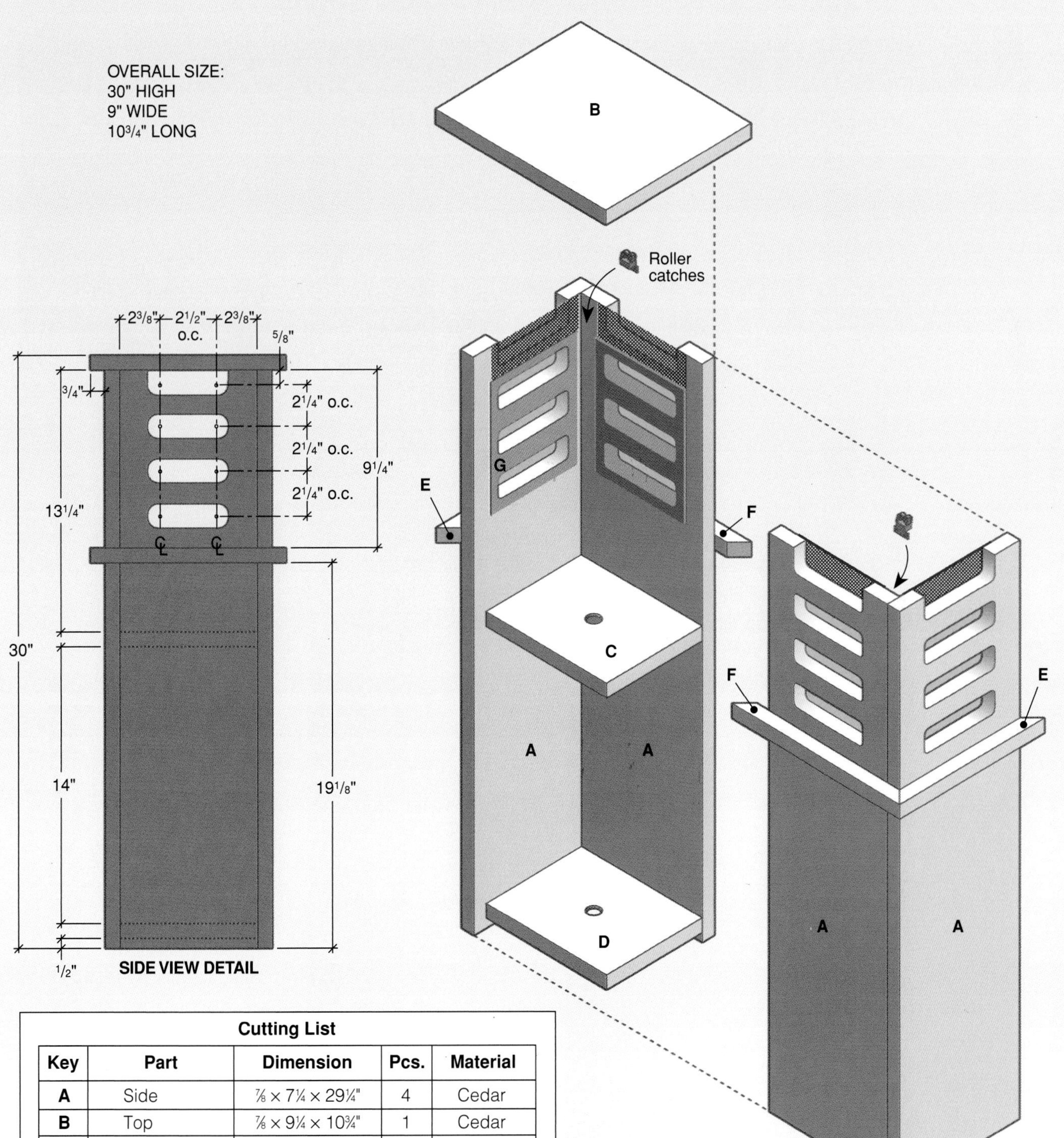


**Cutting List**

| Key | Part | Dimension | Pcs. | Material |
|---|---|---|---|---|
| A | Side | ⅞ × 7¼ × 29¼" | 4 | Cedar |
| B | Top | ⅞ × 9¼ × 10¾" | 1 | Cedar |
| C | Shelf | ⅞ × 7¼ × 5½" | 1 | Cedar |
| D | Bottom | ⅞ × 7¼ × 5½" | 1 | Cedar |
| E | Short trim | ⅞ × ⅞ × 8⅞" | 2 | Cedar |
| F | Long trim | ⅞ × ⅞ × 10⅝" | 2 | Cedar |
| G | Diffuser | ¹⁄₁₆ × 5¼ × 6¾" | 4 | Acrylic |

**Materials:** Fiberglass screen fabric, 1⅝" yellow deck screws, 4d finish nails, ½" self-tapping pan-head screws, ⁵⁄₁₆" staples, surface-mounted outdoor light fixture, type UF exterior electrical cable, roller catches (2).

**Note:** Measurements reflect the actual thickness of dimensional lumber.

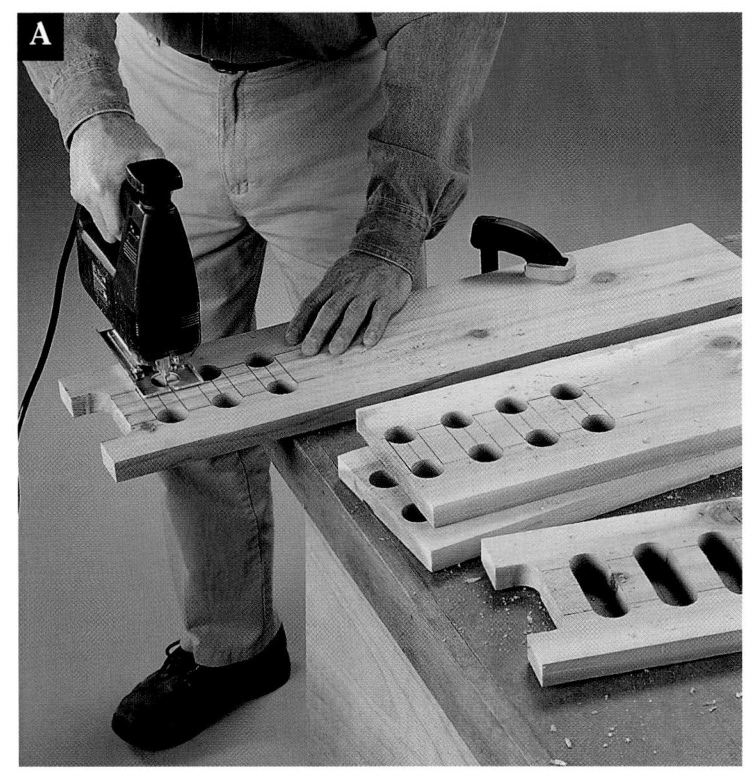

*Create the diffuser and screen openings by drilling the ends with a 1¼" spade bit and completing the cuts with your jig saw.*

*Attach the diffusers with pan-head screws, taking care not to fracture the acrylic.*

## Directions: Exterior Light Post

MAKE THE SIDES. Cedar lumber typically comes ⅞" thick, with one side planed smooth and the other side left rough. For this project we turned the smooth side out.

Start by cutting the sides (A) to length from 1 × 8 boards. Mark the centerpoints for the holes that will become the openings for the diffusers and screens (see *Diagram*). Drill the holes with a 1¼" spade bit. Draw lines to connect the pairs of holes, and use your jig saw to cut along the lines and complete the openings **(photo A).** Sand the insides of the openings with a 1" drum sander mounted on your drill; a light touch here will yield best results because of the softness of cedar wood.

To cut the diffusers, mark cutting lines on acrylic, then score the lines using a utility knife and metal straightedge. Then, snap the acrylic at the lines.

Position a diffuser over the lower three openings on each side piece and drill screw holes very slowly with a 1⁄16" bit (a special bit for drilling plastic is recommended but not essential). Mount each diffuser with ½" self-tapping pan-head screws **(photo B).** After the diffusers are attached, cut the screens to size and fasten over the top openings with 5⁄16" staples **(photo C).**

Mark the locations of the shelf and bottom on the inner faces of the four sides **(photo D).** Flip the sides over and measure down from the top edges to mark the locations of the trim pieces on the outer faces.

Next, drill countersunk pilot holes along both edges of two sides. Fasten three sides together with deck screws.

CUT AND INSTALL THE SHELF, BOTTOM AND TOP. Double-check the actual dimensions of your project, then cut the shelf (C) and bottom (D) to size. Using a spade bit, drill a ¾" hole for the electrical cable in the center of both pieces. Position the shelf and bottom in place, drill countersunk pilot holes and secure with deck screws. Attach a generous length (3 to 4') of UF exterior electrical cable to the fixture and mount the fixture in the center of the shelf. Feed the wire through the bottom **(photo E).**

Cut the top (B) to size from 1 × 10 stock, and center it on the partially assembled light post. Position and install the roller catches.

**TIP**

*Cedar is a soft wood with large annual rings. It expands and contracts in response to changes in humidity more dramatically than many other woods. Consider buying cedar lumber several days in advance so it can stabilize in its new environment.*

*Stretch the screens tight across the top openings and attach in place with 5⁄16" staples.*

*Measure up from the bottom edge to mark positions of the shelf, bottom and trim pieces.*

COMPLETE THE ASSEMBLY. Screw the remaining side in place. Cut the trim pieces (E, F), mitering the corners. Drill pilot holes and attach the trim pieces to the light post with 4d galvanized finish nails.

APPLY THE FINISHING TOUCHES. Sand the surfaces and break the edges with medium-grit sandpaper. One of the benefits of cedar for outdoor projects is that it doesn't require a finish. It will, however, quickly develop a weathered appearance unless an exterior stain or clear exterior finish is applied.

INSTALLATION. Unless you have considerable experience with electrical projects, hire an electrician to run wiring to your light post. If you decide to do this work yourself, make sure to check on local Code restrictions. One option for mounting the light post in your yard is to attach a ¾" pipe flange and an 18" length of galvanized pipe to the bottom of the light post, and embed the pipe in firmly packed earth or a concrete footing.

*Pull the electrical cable from the fixture through the ¾" hole in the bottom of the light post.*

PROJECT
POWER TOOLS

# Window Seat

*Curl up with a good book,
or just enjoy the view from this cozy window seat.*

CONSTRUCTION MATERIALS

| Quantity | Lumber |
|---|---|
| 3 | 1 × 2" × 8' oak |
| 1 | 1 × 2" × 6' oak |
| 1 | 1 × 3" × 6' oak |
| 4 | 1 × 4" × 6' oak |
| 9 | ½ × 1¾" × 4' oak* |
| 1 | ½ × 2¾" × 2' oak* |
| 8 | ½ × 2¾" × 3' oak* |
| 2 | ½ × 2¾" × 4' oak* |
| 1 | ½ × 2¾" × 5' oak* |
| 6 | ½ × 3¾" × 5' oak* |
| 1 | ¾" × 2 × 6' oak plywood |

*Stock sizes commonly available at most woodworking specialty stores.

You'll find this Mission-style window seat to be an excellent place to spend an afternoon. Though it fits nicely under a window, the frame is wide enough so you won't ever feel cramped. The length is perfect for taking a nap, enjoying a sunset or watching children playing in the yard. Or perhaps you'd prefer to sit elsewhere to simply admire your craftsmanship from a distance.

Our project uses oak for its strength and warm texture, and includes a frame face and nosing trim for a more elegant appearance. The rails are capped to make comfortable armrests, and the back is set lower than the sides so it won't block your window view. Though this project has many parts, it requires few tools and is remarkably easy to build. A few hours of labor will reward you with a delightful place to enjoy many hours of relaxation.

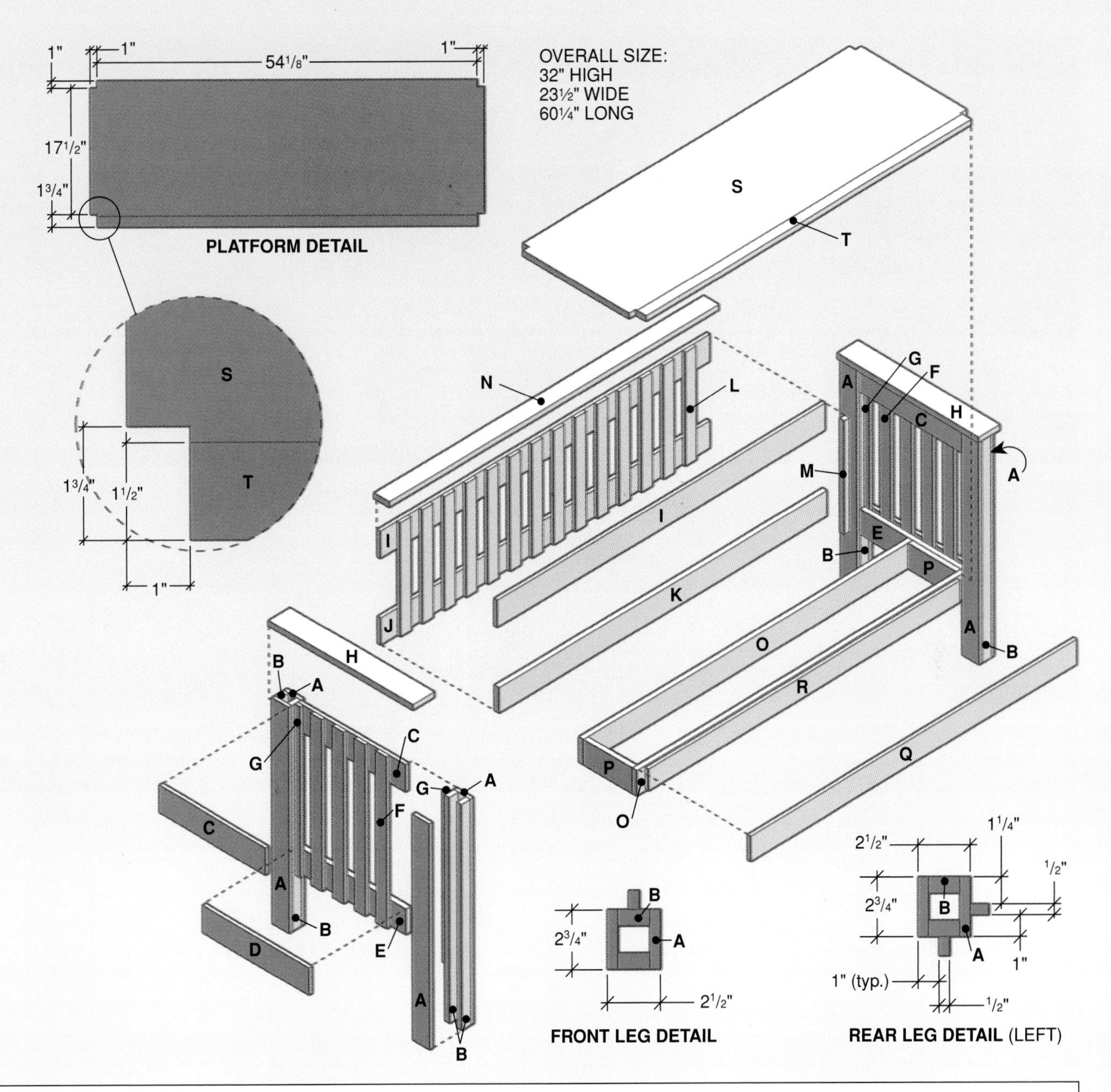


### Cutting List

| Key | Part | Dimension | Pcs. | Material |
|---|---|---|---|---|
| A | Wide leg piece | ½ × 2¾ × 31¼" | 8 | Oak |
| B | Narrow leg piece | ¾ × 1½ × 31¼" | 8 | Oak |
| C | End top rail | ½ × 3¾ × 17½" | 4 | Oak |
| D | Outer bottom rail | ½ × 3¾ × 17½" | 2 | Oak |
| E | Inner bottom rail | ¾ × 3½ × 17½" | 2 | Oak |
| F | End slat | ½ × 1¾ × 23¾" | 8 | Oak |
| G | End half slat | ½ × ⅞ × 23¾" | 4 | Oak |
| H | End cap | ¾ × 3½ × 23½" | 2 | Oak |
| I | Back top rail | ½ × 3¾ × 54¼" | 2 | Oak |
| J | Outer bottom rail | ½ × 3¾ × 54¼" | 1 | Oak |

### Cutting List

| Key | Part | Dimension | Pcs. | Material |
|---|---|---|---|---|
| K | Inner bottom rail | ¾ × 3½ × 54¼" | 1 | Oak |
| L | Back slat | ½ × 1¾ × 15¾" | 14 | Oak |
| M | Back half slat | ½ × ⅞ × 15¾" | 2 | Oak |
| N | Back cap | ¾ × 2½ × 54¼" | 1 | Oak |
| O | Support side | ¾ × 3½ × 54¾" | 2 | Oak |
| P | Support end | ¾ × 3½ × 8" | 2 | Oak |
| Q | Frame face | ½ × 3¾ × 54¼" | 1 | Oak |
| R | Spacer | ½ × 2¾ × 52" | 1 | Oak |
| S | Platform | ¾ × 18¾ × 56⅛" | 1 | Oak Ply. |
| T | Platform nosing | ¾ × 1½ × 54⅛" | 1 | Oak |

**Materials:** Wood glue, 1" brads, wood screws (⅝",1¼", 1½"), 4d finish nails, oak-veneer edge tape (8'), finishing materials.

**Note:** Measurements reflect the actual thickness of dimensional lumber.

*Assemble the legs with glue and clamps, using wax paper to protect your worksurface.*

*Attach the end slats to the outer rails with glue and wood screws, using a spacer as a guide.*

## Directions: Window Seat

ASSEMBLE THE LEGS. Each leg piece consists of four pieces glued together.

Cut the wide leg pieces (A) and narrow leg pieces (B) to size, and sand the cut edges smooth. For each leg, lay a narrow leg piece on your worksurface, then butt a wide leg piece against an edge to form an "L." Apply wood glue and clamp the pieces together **(photo A).** Assemble and glue together another "L" in the same fashion. Then, glue the two L-assemblies together to form a leg. Repeat this process to make the other legs.

BUILD THE END ASSEMBLIES. To ensure that the end rails and slats remain square during the assembly process, build a simple jig by attaching two 2 × 2" boards at a 90° angle along adjacent edges of a 24 × 48" piece of plywood.

Begin by cutting the end top rails (C), outer bottom rails (D), inner bottom rails (E) and end slats (F) to size, and sand the edges smooth.

Place a top rail and an outer bottom rail in the jig. Position a slat over the rails, 2⅜" in from the ends. Adjust the pieces so ends of the slat are flush with the edges of the rails, and keep the entire assembly tight against the jig. Attach with glue and ⅝" wood screws driven through countersunk pilot holes.

Using a 1¾"-wide spacer, attach the remaining end slats with glue and ⅝" screws **(photo B).** NOTE: Make sure to test-fit all the slats for uniform spacing before attaching them to the rails.

Now, position a bottom rail over the slats, ¼" up from the bottom edges of the slats, and attach with glue and countersunk 1¼" screws. Place a top rail over the slats and attach with glue and 1" brads.

> **TIP**
>
> *Take care to countersink all screw heads completely when building furniture that will be used as seating.*

*Attach the end half slats to the legs with glue and countersunk wood screws.*

*Attach the lower inner rail with glue and countersunk screws; the upper inner rail with glue and finish nails.*

Repeat the process to build the other end assembly.

BUILD THE BACK ASSEMBLY. The back is constructed in a similar manner to that used for the end assemblies. Again, use the jig to keep the back assembly square.

Cut the back top rails (I), the outer bottom rail (J), the inner bottom rail (K) and the back slats (L) to length and sand the cuts smooth. Place a top rail and the inner bottom rail in the jig. Place a back slat on the rails, 2⅝" in from the ends. Adjust the pieces so the ends of the slat are flush with the edge of the top rail and overhang the edge of the bottom rail by ¼". Attach the slat with glue and ⅝" wood screws driven through countersunk pilot holes.

Test-fit the remaining slats, using a spacer as a guide, then attach with glue and ⅝" wood screws. Position the remaining bottom rail so the edge is flush with the bottom edges of the slats, and attach with glue and 1" brads. Place the remaining top rail over the slats and attach it with glue and 1" brads.

JOIN THE LEGS TO THE END ASSEMBLIES. Half slats attached to the legs will complete the slat pattern and serve as cleats for attaching the end assemblies.

Cut the end half slats (G) to size from ½ × 2¾" × 4' stock and sand smooth. Place each leg on your worksurface with a narrow leg piece facing up. Center the half slat on the face of the leg (see *Diagram*), with the top ends flush. Drill countersunk pilot holes in the half slat, locating them so the screw heads will be covered by the rails when the seat is completed, then attach the half slats to the legs with glue and 1¼" screws **(photo C).**

Position an end assembly between a front and rear leg so the half slats fit between the rails and the top edges are flush. Drill counterbored pilot holes through the inner bottom rail and into the half slats, taking care to avoid other screws, then attach with glue and 1¼" screws. Attach the top rail to the half slats with glue and 4d finish nails driven through pilot holes **(photo D).** Repeat this process for the other end assembly.

MAKE THE SUPPORT FRAME. The support frame is attached to the inner bottom rails on the end assemblies, and will support the seat.

Cut the support sides (O) and ends (P) to length, and

*Attach the support frame with glue and countersunk screws driven through the support end and into the inner bottom rail.*

*Glue the platform nosing to the platform and hold it in place with bar clamps.*

sand the cuts smooth. Position the ends between the sides (see *Diagram),* then drill countersunk pilot holes and join the pieces with glue and 1¼" screws.

Lay one end assembly on your worksurface, and position the support frame upright so the front corner of the frame is tight against the front leg and the edges of the frame are flush with the edges of the bottom rail. Drill counterbored pilot holes and attach the support frame to the end assembly with glue and 1¼" screws **(photo E).** Stand the window seat upright, and clamp the other end in position. Drill counterbored pilot holes and attach with 1¼" screws.

ATTACH THE BACK. Like the end assemblies, the back assembly is joined to the legs with half slats.

Cut the back half slats (M) to size from ½ × 2¾" × 2' stock and sand smooth. On the inside face of each rear leg, measure 7½" up from the bottom, and draw a horizontal line at this point. Measure in 1¼" from the back edge of the leg along this line, and draw a vertical line upward.

Position a half slat against the leg so its rear edge is on the vertical line and its bottom edge is on the horizontal line. Drill countersunk pilot holes and attach the half slat to the leg with glue and 1½" screws. Repeat with the other rear leg.

After the half slats have been attached to each rear leg, slide the back assembly over the half slats so the top edges are flush. Drill countersunk pilot holes through the inner bottom rail into the half slat and attach with glue and 1¼" screws. Drill pilot holes and join the top rail to the back half slats with glue and 4d finish nails.

ATTACH THE CAPS. Caps are attached to the ends and back of the window seat to create armrests and backrests.

Cut the end caps (H) and back cap (N) to length. Center the end caps over the end assemblies, with the back edges flush. Drill counterbored pilot holes through the end caps and into the legs, and attach with glue and 1½" screws.

Position the back cap over the back assembly so the front edge is flush with the front edges of the legs. Drill counterbored pilot holes through the back cap into the top rails. Attach with glue and 1½" screws.

MAKE THE PLATFORM. Because the platform is made of plywood, the edges must be covered with oak nosing and

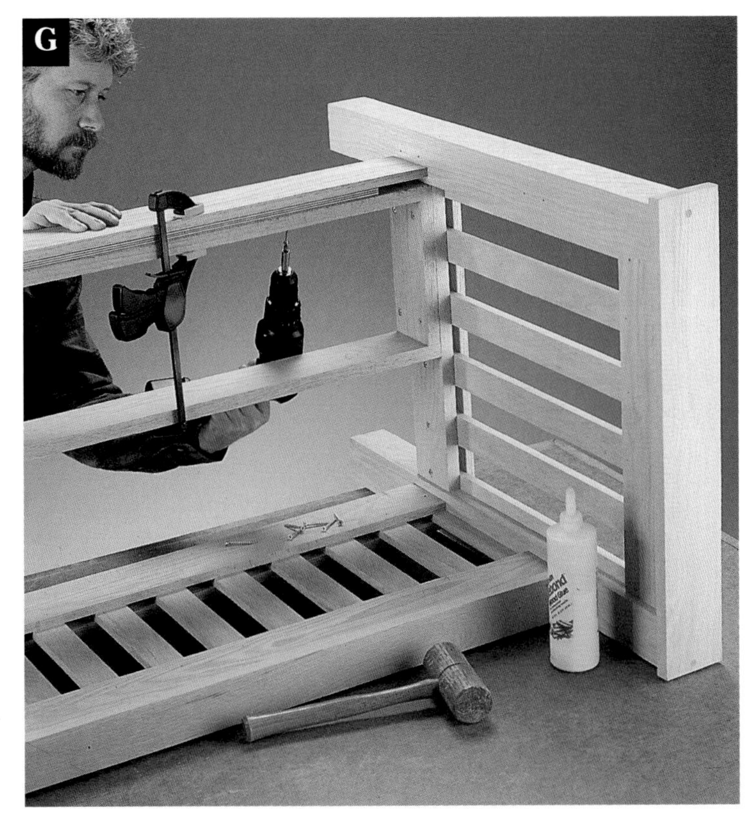

*Clamp the spacer and frame face to the support frame and attach with glue and screws driven through the inside of the support frame.*

*Attach the platform to the rails, frame face and support frame with glue and counterbored screws.*

edge tape to create the appearance of solid wood.

Cut the platform (S) and platform nosing (T) to size, and sand smooth. Glue the nosing to the front edge of the platform, leaving 1" exposed on each end, and clamp in place until the glue dries **(photo F).**

To accommodate the legs, use a jig saw to cut a 1 × 1" notch in each back corner of the platform and a 1 × 1¾" notch in each front corner (see *Diagram*). Apply self-adhesive oak veneer edge tape to the side and back edges of the platform (don't apply tape to the notches). Lightly sand the edges of the tape until they are smooth.

ATTACH THE FRAME FACE. The frame face and spacer are attached to support the front edge of the platform and create design consistency.

First, cut the frame face (Q) and spacer (R) to length, and sand the cuts smooth. Use glue to join the pieces together, centering the spacer on the frame face. Clamp the pieces together until the glue dries.

Position the frame face assembly against the front of the support frame so the top edges of the face and support frame are flush. Drill countersunk pilot holes from inside the support frame, then attach with glue and 1¼" screws **(photo G).**

ATTACH THE PLATFORM. Drill counterbored pilot holes and attach the platform to the support the frame, frame face and bottom rails with glue and 1½" screws **(photo H).**

APPLY FINISHING TOUCHES. Plug the counterbored holes with glued oak plugs and fill all visible nail holes with putty. Scrape off any excess glue and finish-sand the window seat. Apply a stain of your choice (ours is medium oak), then add a coat of polyurethane.

Add seat cushions that complement the wood tones of the window seat and the overall decorating scheme of your room.

### TIP

*If you find any nail holes that were not filled before you applied stain and finish, you can go back and fill the holes with a putty stick that closely matches the color of the wood stain.*

PROJECT
POWER TOOLS

# Mirrored Coat Rack

*Nothing welcomes visitors to your home like an elegant, finely crafted mirrored coat rack.*

### Construction Materials

| Quantity | Lumber |
|---|---|
| 1 | 1 × 2" × 3' oak |
| 1 | 1 × 3" × 4' oak |
| 1 | 1 × 4" × 3' oak |
| 1 | 1 × 6" × 3' oak |
| 1 | ½ × ¾" × 4' egg-and-dart molding |
| 1 | ¼" × 2 × 4' plywood |

An entryway or foyer seems naked without a coat rack and a mirror, and this simple oak project gives you both features in one striking package. The egg-and-dart beading at the top and the decorative porcelain and brass coat hooks provide just enough design interest to make the project elegant without overwhelming the essential simplicity of the look.

We used inexpensive red oak to build our mirrored coat rack, but if you are willing to invest a little more money, use quartersawn white oak to create an item with the look of a true antique. For a special touch, have the edges of the mirror beveled at the glass store.

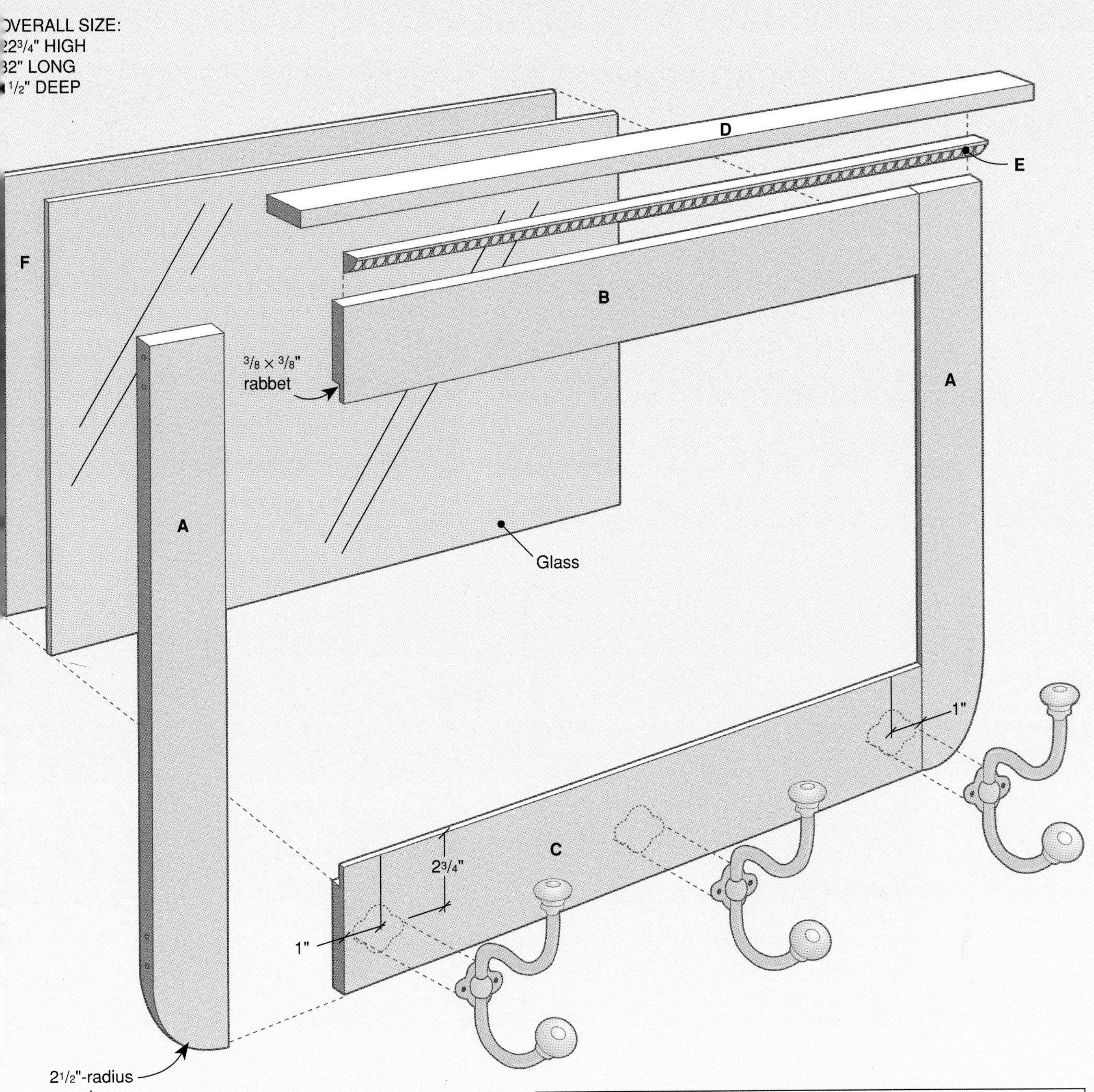


| Key | Part | Dimension | Pcs. | Material |
|---|---|---|---|---|
| A | Stile | ¾ × 2½ × 22" | 2 | Oak |
| B | Top rail | ¾ × 3½ × 24" | 1 | Oak |
| C | Bottom rail | ¾ × 5½ × 24" | 1 | Oak |
| D | Cap | ¾ × 1½ × 32" | 1 | Oak |
| E | Molding | ½ × ¾ × 29" | 1 | Oak |
| F | Mirror back | ¼ × 13¾ × 24¾" | 1 | Plywood |

**Materials:** ⅛ × 13¾ × 24¾" mirror, wood glue, ¼ × 36" oak doweling, #6 × 1½" wood screws, coat hooks with screws (3), 1" wire brads.

**Note:** Measurements reflect the actual thickness of dimensional lumber.

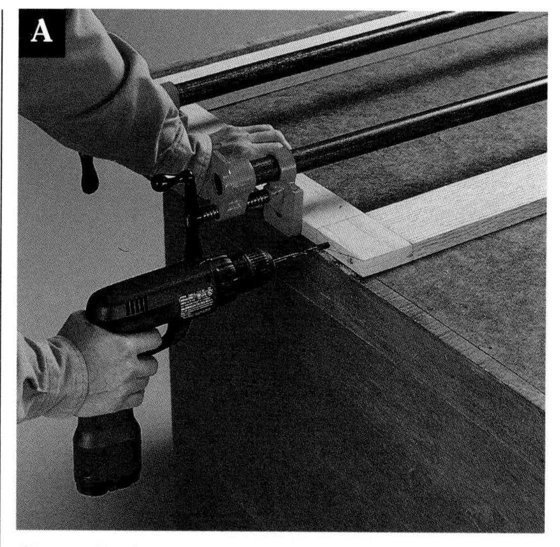


*Clamp the frame components together, then drill 3½"-deep guide holes for the through-dowel joints.*

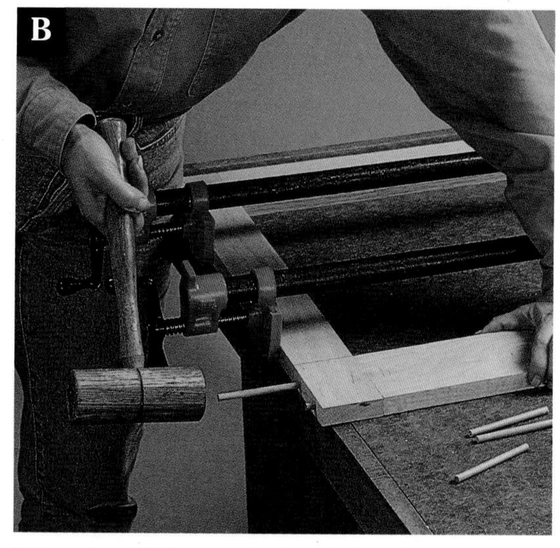


*Drive glued 4"-long oak dowels into the guide holes to make the dowel joints.*

*Mount a belt sander to your worksurface, and use it as a grinder to smooth out the roundover cuts on the frame.*

## Directions: Mirrored Coat Rack

MAKE THE MIRROR FRAME. Start by cutting the frame components to size. Cut the stiles (A) to length from 1 × 3 oak. Cut the top rail (B) from 1 × 4 oak and cut the bottom rail (C) from 1 × 6 oak. Sand the stiles and rails with medium sandpaper (100- or 120-grit) to smooth out rough spots, then finish-sand with fine sandpaper (150- or 180-grit). We used through-dowel joints to hold the frame parts together.

Lay the rails between the stiles on your worksurface to form a frame. Square the frame, then use pipe or bar clamps to hold it together. Drill two ¼"-dia. × 3½"-deep holes at each joint, drilling through the stiles and into the rails **(photo A).** Now, cut eight 4"-long oak dowels. Apply glue and drive dowels into holes, using a wood mallet so you don't break the dowels **(photo B).** Once the glue has dried, remove the clamps, trim off the ends of the dowels with a backsaw, sand them level with the wood surface and scrape off excess glue.

ROUND OVER THE FRAME ENDS. On the bottom end of each stile, lay out an arc with a 2½" radius (see *Diagram*) to mark the decorative roundovers. Cut along the arc line using a jig saw, and smooth out the cut with a belt sander mounted to your worksurface **(photo C).**

DRILL MOUNTING HOLES & CUT THE MIRROR RECESS. Before you attach the decorative top cap, drill ⅜"-diameter counter-

### TIP

*Through-dowel joints are the easiest dowel joints to make—all you need is a good pipe or bar clamp, and the ability to drill a reasonably straight guide hole. The visible dowel ends at the joints usually contribute to the general design of the project.*

*Use a router with a ⅜" piloted rabbeting bit to cut a recess for the mirror in the frame back.*

*Center egg-and-dart trim molding under the cap, and attach with glue and 2d finish nails.*

*Install the mirror and mirror back, then secure them to the frame with wire brads, driven with a brad pusher.*

bored screw holes through the top rail, so you can attach the mirrored coat rack to a wall. Next, cut a rabbet around the back edge of the inside frame, using a router and ⅜" piloted rabbet bit. Set the depth of the cut at ½" **(photo D).** Square off the corners of the rabbet with a wood chisel.

INSTALL THE CAP & MOLDING. Cut the cap (D) to length from 1 × 2 oak and attach it to the top of the top rail, flush with the back edge, using glue and counterbored wood screws. Make sure the cap overhangs the stiles evenly on the ends (1½" per end).

Cut a piece of oak egg-and-dart-style trim (E), or any other trim style you prefer, to length. Sand a slight, decorative bevel at each end. Attach the molding tight against the underside of the cap, centered side to side, using glue and 2d finish nails driven with a tack hammer **(photo E).** Set the nail heads. Sand all sharp edges on the frame.

INSTALL THE MIRROR. Set the mirror into the recess in the frame. Cut ¼"-thick plywood to make the mirror back (F), and set it over the back of the mirror. Secure the mirror and mirror back with 1" wire brads driven into the edges of the frame **(photo F).**

TIP

*Try to hit a wall stud with at least one mounting screw when hanging heavy objects on a wall. Use 3"-long screws when attaching to wall studs and use toggle bolts to mount where no studs are present.*

APPLY FINISHING TOUCHES. Fill all counterbores with oak plugs and sand flush with the wood surface. Apply stain and topcoat as desired. When dry, install the coat hooks (see *Diagram*). Hang the coat rack (see *Tip*, above).

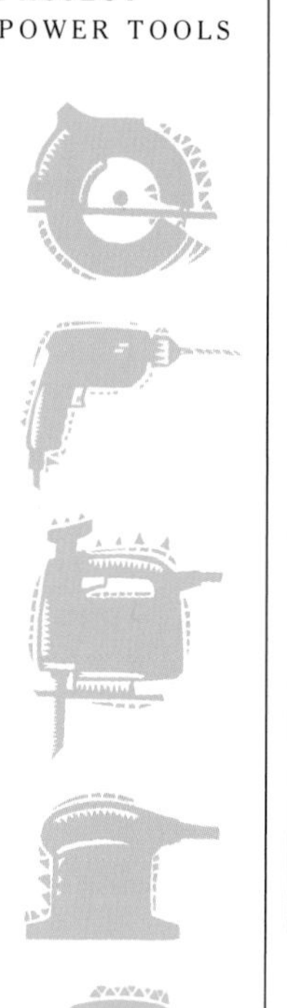

# Jewelry Box

*This piece of fine furniture will be a worthy home for your family treasures.*

**Construction Materials**

| Quantity | Lumber |
|---|---|
| 1 | ¾ × 24 × 48" MDF* |
| 1 | ½ × 12 × 30" birch plywood |
| 1 | ½ × 24 × 30" birch plywood |
| 1 | ¼ × 12 × 24" hardboard |

*Medium-density fiberboard

Without a suitable home, jewelry has a way of getting lost or misplaced. This elegant and roomy jewelry box solves that problem with pizzazz.

Our classically proportioned chest—like all fine furniture—is as functional as it is beautiful. Three spacious drawers accommodate everything from fun and funky costume jewelry to the finest family heirlooms.

The precision craftsmanship utilizes a simple system of dadoes and rabbets to achieve the close tolerances and tight joints which characterize true quality woodwork.

The timeless design of this piece allows for many options in materials and finish, providing great flexibility for customizing your box to suit a special person or unique situation.

Building this project as a gift will showcase your thoughtfulness as well as your woodworking skill.

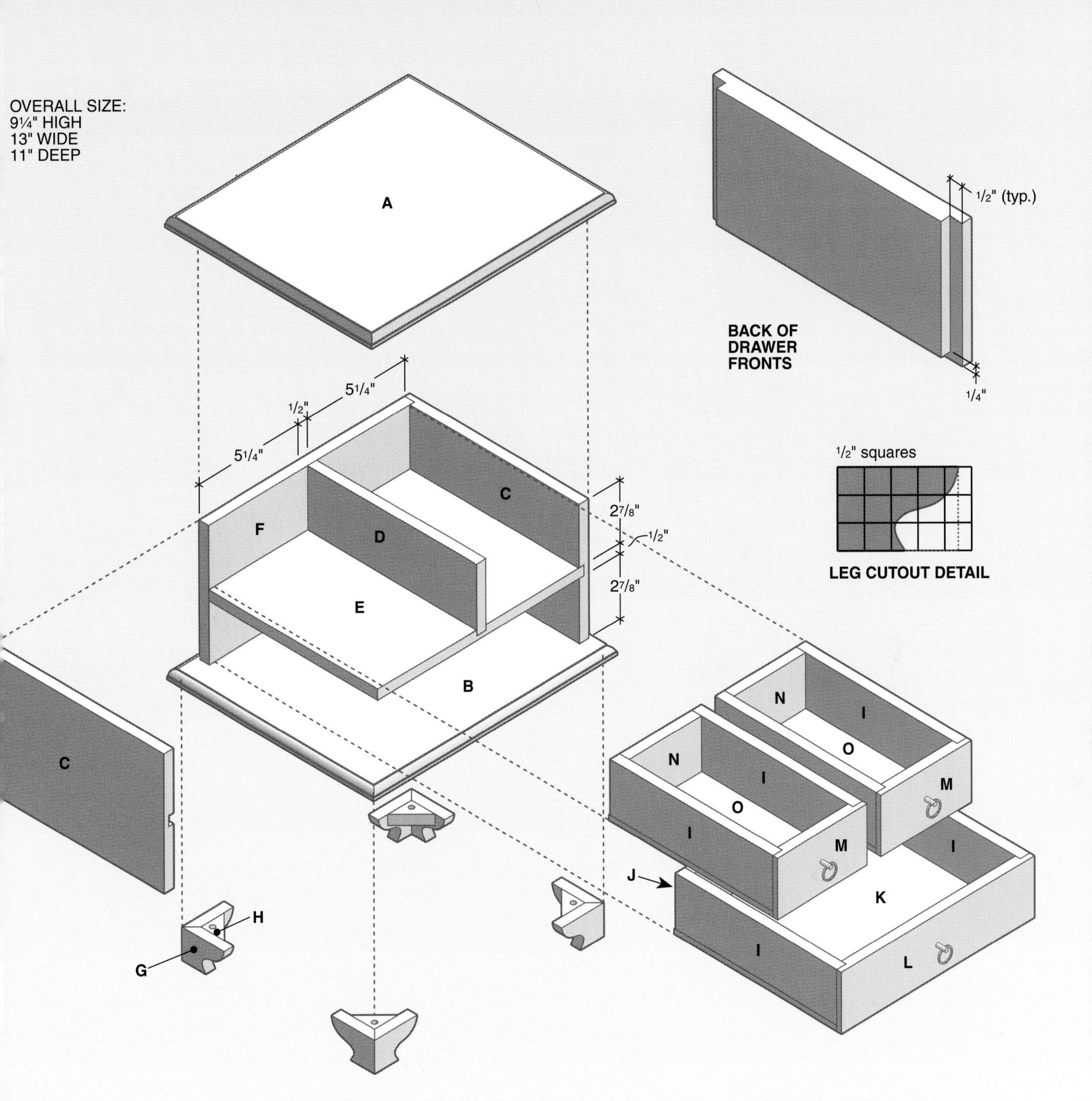


**Cutting List**

| Key | Part | Dimension | Pcs. | Material |
|---|---|---|---|---|
| A | Top | ¾ × 11 × 13" | 1 | MDF |
| B | Bottom | ¾ × 11 × 13" | 1 | MDF |
| C | Side | ½ × 6¼ × 9½" | 2 | Birch ply. |
| D | Divider | ½ × 3⅛ × 9" | 1 | Birch ply. |
| E | Shelf | ½ × 9 × 11" | 1 | Birch ply. |
| F | Back | ½ × 6¼ × 11" | 1 | Birch ply. |
| G | Leg | ½ × 1½ × 2¼" | 8 | Birch ply. |
| H | Glueblock | ½ × 1¼ × 1¼" | 4 | Birch ply. |

**Cutting List**

| Key | Part | Dimension | Pcs. | Material |
|---|---|---|---|---|
| I | Drawer side | ½ × 2½ × 8 11/16" | 6 | Birch ply. |
| J | Long drwr. back | ½ × 2½ × 9⅜" | 1 | Birch ply. |
| K | Long drwr. bottom | ¼ × 8¾ × 10⅜" | 1 | Hardboard |
| L | Long drwr. front | ½ × 2¾ × 10⅜" | 1 | Birch ply. |
| M | Short drwr. front | ½ × 2¾ × 4⅞" | 2 | Birch ply. |
| N | Short drwr. back | ½ × 2½ × 3⅞" | 2 | Birch ply. |
| O | Short drwr. bottom | ¼ × 4⅞ × 8¾" | 2 | Hardboard |
| | | | | |

**Materials:** Wood glue, brads, 4d finish nails, #6 × 1" screws, drawer pulls (3), finishing materials.

**Note:** Measurements reflect the actual thickness of dimensional lumber.

*To cut the shelf dadoes, rout both sides in one pass, using a clamped straightedge as a guide for the router base.*

*Apply glue to the shelf, the divider and the shoulders of the rabbets before attaching the back.*

## Directions: Jewelry Box

CUT AND SHAPE THE CABINET PARTS. Measure and cut the top (A) and bottom (B) from ¾" MDF. Shape the top edges of both pieces using a router with a ⅜" standard roundover bit.

Measure and cut out the sides (C), divider (D), shelf (E) and back (F). Mark the sides and the shelf for location of dadoes (see *Diagram*), and mark the back edges of the sides for the ½ × ¼"-deep rabbets. To cut the dadoes, clamp the side blanks to your worksurface with back edges butted, clamp a straightedge in place to guide the router base **(photo A),** and use a ½" straight router bit set ¼" deep. Cut the divider dado in the shelf and the back rabbets in the sides using the same process. Drill pilot holes in the dadoes and rabbets.

ASSEMBLE THE CABINET. Attach the divider to the shelf with glue and brads. Stand the shelf/divider assembly on end and attach one side with glue and brads; flip the assembly and attach the other side. Stand the partially assembled cabinet on its front and attach the back with glue and brads **(photo B).** Next, attach the top by centering it on the assembly, drilling pilot holes, and fastening with glue and 4d finish nails. Flip the cabinet over and attach the bottom in the same manner.

MAKE AND ATTACH THE LEGS. Cut four blanks, 6" or longer, from ½ × 1½" stock. Transfer the leg profile (see *Diagram*) to the ends of the blanks and cut the profiles with a jig saw. Clamp the blanks together and gang-sand the cut edges with a drum sander mounted in your drill. Using a power miter box, cut the legs (G) to length, mitering the ends at 45° **(photo C).** Make sure to cut four *pairs* rather than eight identical pieces. Cut the glueblocks (H) to size from ½" scrap.

Assemble the legs by gluing them in pairs to the glueblocks

### TIP

*Dadoes and rabbets provide strength by increasing the surface area of glue joints and locking components in their proper positions. When machining is done correctly, assembly is almost foolproof. However, accuracy and precision are critical to success, so care invested "up front" will yield dividends later in the process.*

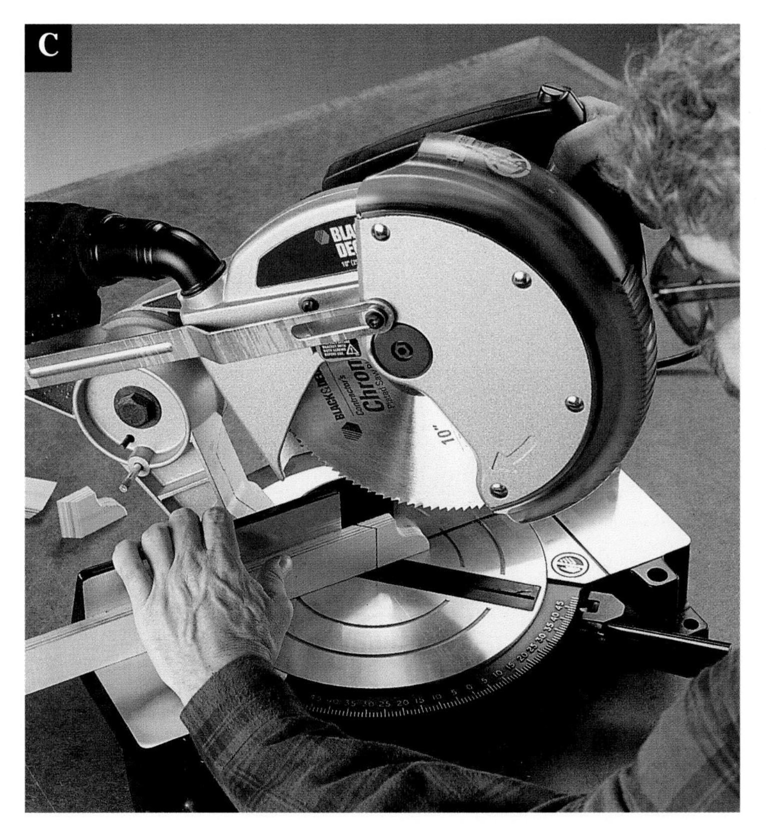


*After the leg blanks have been profiled and sanded, cut the leg pieces to length with a power miter box.*

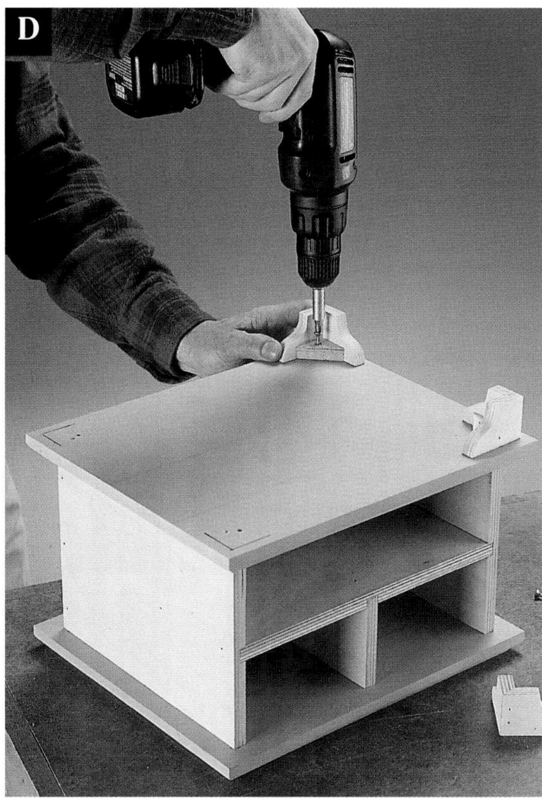


*Attach the legs with screws and glue, leaving a ⅜" overhang along both sides.*

and clamping the pieces with masking tape. After the glue dries, position the legs on the cabinet bottom, drill pilot holes and attach with glue and 1" screws **(photo D).**

BUILD THE DRAWERS. Start by cutting the drawer faces (L, M) to size. Rabbet the backs of these pieces to accept the ½" drawer sides and ¼" bottoms (see *Diagram*). Cut the drawer sides (I) and drawer backs (J, N) to size. Drill pilot holes and assemble the drawer boxes with glue and brads. Cut the drawer bottoms (K) and (O) to size from ¼" hardboard and fasten with glue and brads **(photo E).** Measure and drill for the drawer pulls.

APPLY FINISHING TOUCHES. Set all nail heads, and fill voids with putty. Finish-sand the project, finish as desired and install drawer pulls.

*Attach the drawer bottoms with glue and brads. The bottoms hold the drawers square so they fit within their compartments.*

PROJECT
POWER TOOLS

# Game Table

*Kids will appreciate the size of this table, but the adults will remember the craftsmanship.*

CONSTRUCTION MATERIALS

| Quantity | Lumber |
|---|---|
| 3 | 1 × 2" × 8' oak |
| 3 | 1 × 3" × 8' oak |
| 1 | ¾" × 2 × 2' MDF* |
| 1 | ¼" × 2 × 4' hardboard |

* Medium-density fiberboard

Our game table is a striking piece of furniture that can easily double as a decorative end table. Sturdy leg units and an internal cleat structure make this table durable and stable enough for constant use. A slim drawer slides underneath the game table, great for holding score pads, cards or other gaming supplies.

The most striking feature of our game table is the veneered top. The pattern is accomplished with four panels of self-adhesive oak veneer applied in different directions for a unique appearance. Rabbet joints cut with a router join the oak border trim to the tabletop.

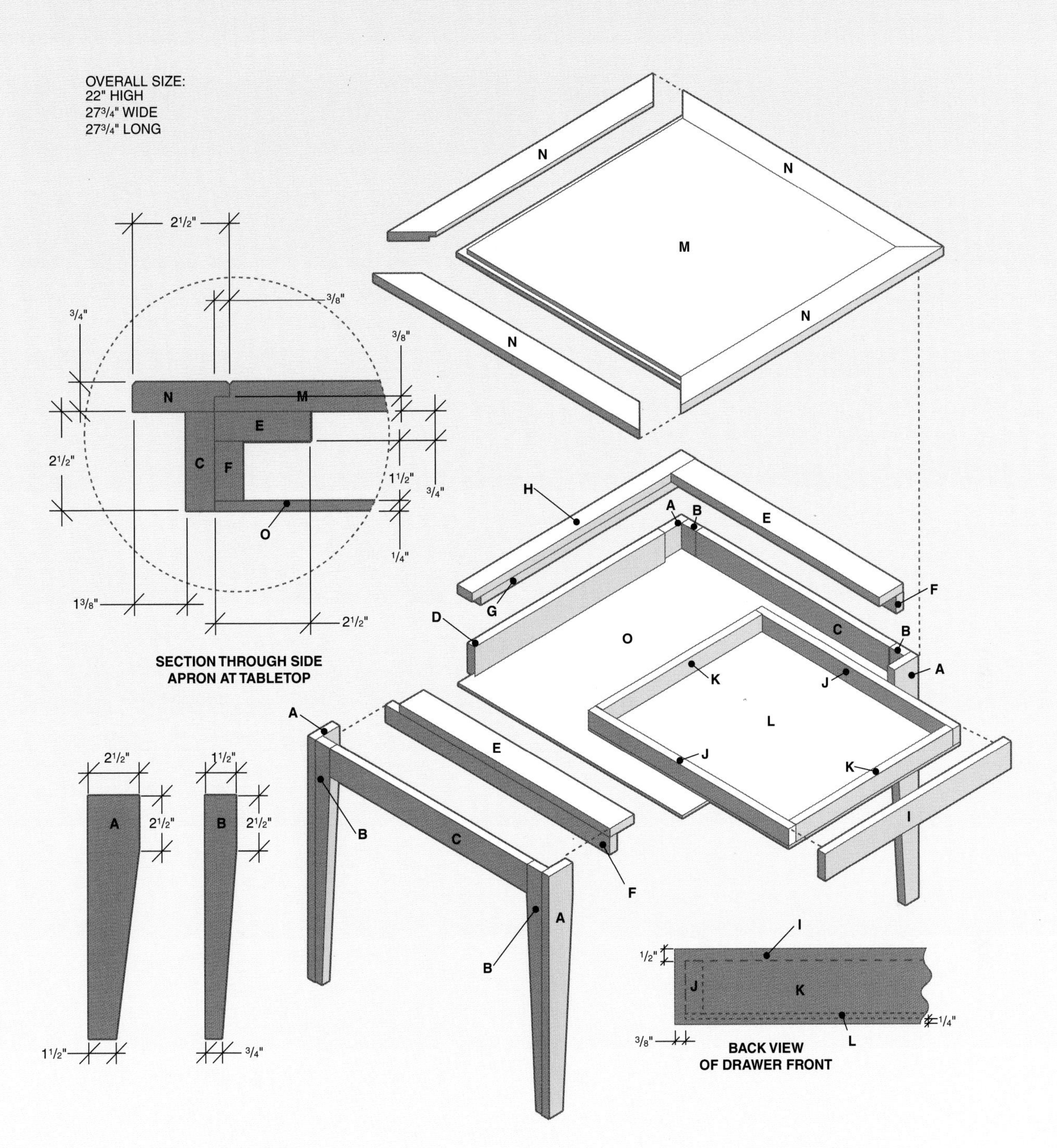


| Key | Part | Dimension | Pcs. | Material |
|---|---|---|---|---|
| A | Wide leg piece | ¾ × 2½ × 21¼" | 4 | Oak |
| B | Narrow leg piece | ¾ × 1½ × 21¼" | 4 | Oak |
| C | Side apron | ¾ × 2½ × 20½" | 2 | Oak |
| D | Back apron | ¾ × 2½ × 20" | 1 | Oak |
| E | Side stretcher | ¾ × 2½ × 22" | 2 | Oak |
| F | Side cleat | ¾ × 1½ × 23½" | 2 | Oak |
| G | Back cleat | ¾ × 1½ × 22" | 1 | Oak |
| H | Back stretcher | ¾ × 1½ × 23½" | 1 | Oak |

**Cutting List**

| Key | Part | Dimension | Pcs. | Material |
|---|---|---|---|---|
| I | Drawer front | ¾ × 2½ × 19¾" | 1 | Oak |
| J | Drawer side | ¾ × 1½ × 21" | 2 | Oak |
| K | Drawer end | ¾ × 1½ × 16¾" | 2 | Oak |
| L | Drawer bottom | ¼ × 18¼ × 21" | 1 | Hardboard |
| M | Top | ¾ × 23½ × 23½" | 1 | MDF |
| N | Trim | ¾ × 2½ × 27¾" | 4 | Oak |
| O | Bottom | ¼ × 23½ × 23½" | 1 | Hardboard |
| | | | | |

**Materials:** Wood glue, #6 × 1¼" wood screws, 4d finish nails, 2 x 3' self-adhesive oak veneer (2), finishing materials.

**Note:** Measurements reflect the actual thickness of dimensional lumber.

*Use a sharp blade to cut the veneer pieces along the template. Press firmly to prevent the veneer from sliding.*

*Clamp the top securely to your worksurface to keep it from slipping when rabbeting the top.*

## Directions: Game Table

CUT AND ASSEMBLE THE VENEER. The veneer top of the table consists of four triangular quadrants, requiring two large pieces of veneer. Two quadrants are cut from each piece.

Begin by making an 18 × 18" square template from ¼" scrap hardboard. From one corner of the template, measure 17" along adjoining sides and mark points. Tape the bottom edges of the template to your cutting surface. Slide the sheet of veneer under the template until the sides meet the marks and the veneer is positioned at a 45° angle to the edges of the hardboard. Cut with a utility knife down to the marks **(photo A),** then remove the template and cut between the marks to complete the triangle. Repeat the process to cut the three remaining veneer pieces, always keeping the grain aligned as in photo A. Tape the triangular veneer pieces together on their finished faces so the top points meet.

MAKE THE TOP. Cut the top (M) from MDF. Make sure it is perfectly square. Draw diagonal reference lines connecting opposite corners.

Place the veneer on the top, aligning the veneer seams with the diagonal reference lines. When the veneer is positioned correctly, clamp one side in place, then fold back the opposite half and peel away the backing. Lay the exposed veneer back onto the top, and press to bond. Remove the clamps, fold back the remaining half, peel the backing and apply the veneer. Press all seams firmly in place with a J-roller, and trim the edges.

Rabbet the edge of the top, using your router and a ⅜" self-guiding rabbet bit set to a depth of ⅜" **(photo B).** Also rabbet one edge of the 1 × 3 stock for the trim pieces. Before rabbeting this material, use scrap wood to test-fit the depth of the cut against the rabbeted edges of the top to ensure that the faces will be flush.

Using a block plane or sander, make a 1/16" chamfer (bevel) on the perimeter of the top and along both upper edges of the 1 × 3. Cut the trim pieces (N) to length, mitering the ends at 45° angles. Attach the trim pieces to the top with glue, and clamp in place. Drill pilot holes and use finish nails to lock-nail the trim pieces together at the joints. Set the nail heads and fill with putty.

MAKE THE LEGS. Cut the wide and narrow leg pieces (A, B) to length. Glue and clamp the edge of each narrow leg piece to the face of a wide piece so the edges are flush **(photo C).** Once dry, mark the diagonal taper on each leg (see *Diagram*), cut with a jig saw and sand smooth.

BUILD THE CLEAT ASSEMBLIES. Cut the side stretchers (E), side cleats (F), back stretcher (H) and back cleat (G) to length. Arrange the side stretchers and cleats in pairs (see *Diagram*). Drill countersunk pilot holes and attach the stretchers to the cleats with glue and screws. Make sure the back cleat is centered on its stretcher, with a ¾" space at each end.

Join the wide and narrow leg pieces with glue; if this is done carefully, no nails or screws are required.

*Use glue and screws to join the leg pairs to the cleat assemblies and aprons.*

*Fasten the top with glue and screws driven through the stretchers and into the top.*

ASSEMBLE THE TABLE FRAMEWORK. Cut the side aprons (C) and back apron (D) to length. Lay a side apron on your worksurface, and place the right cleat assembly over it so the side stretcher is standing on edge, flush with the top edge of the apron. Arrange the right front and back legs in their correct position, with the narrow leg pieces flat on the worksurface. The side stretcher should butt against the wide leg pieces, and the side apron should butt against the narrow leg pieces. Make sure the parts are flush on the face and the top edge, then glue and clamp the parts together. Fasten with screws driven through the side cleat into the apron and legs **(photo D).** Repeat the process to assemble the left leg pair.

Position the finished leg pairs upright with the back legs resting on the worksurface. Position the back apron and back cleat assembly between the back legs. Adjust so the edges are flush, and fasten with glue and screws.

ATTACH THE TOP AND BOTTOM. Place the top upside down on a clean worksurface. Set the leg/apron assembly over it, aligning the leg corners on the miter joints. Drill pilot holes and attach the top with glue and screws **(photo E).** Cut the bottom (O) to size, drill pilot holes, and fasten to the bottom edges of the cleats using glue and screws.

BUILD THE DRAWER. Cut all drawer parts (I, J, K, L) to size. Place the drawer ends between the drawer sides so the edges are flush. Drill countersunk pilot holes and attach using glue and screws.

Position the drawer bottom on the drawer box, drill pilot holes, and attach with glue and screws. Lay the drawer front facedown on your worksurface and position the drawer box on it (see *Diagram*). Drill pilot holes, and fasten with glue and screws driven through the end and into the front.

APPLY FINISHING TOUCHES. A water-based finish may loosen the veneer adhesive, so use alkyd-based polyurethane instead. The chamfered joint between the veneer and the top frame may be painted black after the finish has cured.

# Umbrella Stand

*Keep your umbrellas, canes and walking sticks at easy reach with this classic umbrella stand.*

PROJECT POWER TOOLS

CONSTRUCTION MATERIALS

| Quantity | Lumber |
|---|---|
| 2 | 1 × 10" × 6' oak |
| 1 | ¾ × ¾" × 4' oak cove |
| 2 | 8 × 22" tin |

This umbrella stand is the perfect rainy-day project. It easily holds up to six umbrellas, canes or walking sticks, so you'll never again have to search for these items on the way out the door. The umbrella stand is a natural in a hallway, entry or foyer and a classy alternative to storing umbrellas haphazardly in your closet.

Built from solid oak for sturdiness and appearance, the umbrella stand has miter-cut top trim and cove molding, and decorative diamond cutouts backed with tin panels. This project can be painted or finished with natural stain. We painted the sides, but used an oak finish on the feet and trim—which allows the piece to blend in nicely with wood staircases and doors.

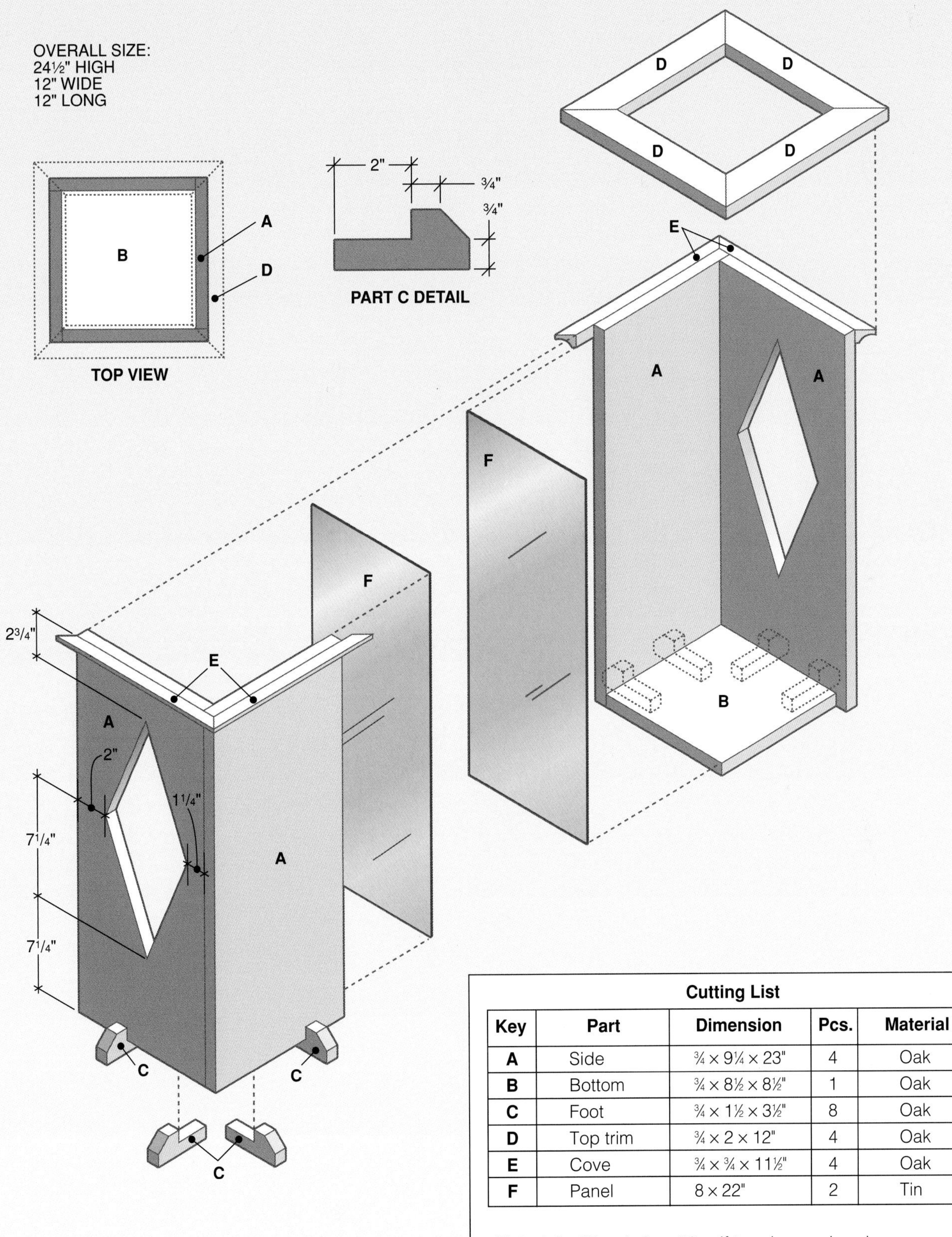


| Key | Part | Dimension | Pcs. | Material |
|---|---|---|---|---|
| A | Side | ¾ × 9¼ × 23" | 4 | Oak |
| B | Bottom | ¾ × 8½ × 8½" | 1 | Oak |
| C | Foot | ¾ × 1½ × 3½" | 8 | Oak |
| D | Top trim | ¾ × 2 × 12" | 4 | Oak |
| E | Cove | ¾ × ¾ × 11½" | 4 | Oak |
| F | Panel | 8 × 22" | 2 | Tin |

**Cutting List**

**Materials:** Wood glue, #6 self-tapping panhead screws, wood screws (#6 × 1¼"), 16-ga. × 1" brads, finish nails (2d and 4d), finishing materials.

**Note:** Measurements reflect the actual thickness of dimensional lumber.

*Drill pilot holes, then cut out the diamond shapes with a jig saw.*

*Use the cutout diamond as a guide when attaching the tin panel around the cutout.*

## Directions: Umbrella Stand

CUT THE SIDES AND BOTTOM. The four sides of the umbrella stand are identical in size, but two of them have decorative diamond cutouts.

Cut the sides (A) and bottom (B) to size. Sand the cuts smooth with medium-grit sandpaper. Draw the diamond on two sides (for measurements, see *Diagram*). NOTE: When the box is assembled, the diamonds will be centered side to side. Drill access holes and use a jig saw to cut out the diamond shapes **(photo A).** Sand the cutouts smooth.

Cut the tin panels (F) to size using aviation snips. Position the tin panels on the inside face of each cutout side, leaving a space at the bottom and along the right edge for the bottom and adjoining sides. Drill pilot holes and attach the tin panel with self-tapping panhead screws. Drive screws at the corners and along the edges of the cutout area. Use a diamond cutout section as a reference when positioning the screws **(photo B).**

*Attach the cove molding to the sides with glue and 1" brads.*

ASSEMBLE THE BOX. The sides are butted together so that one edge of each piece is left exposed (see *Diagram*).

Lay one of the plain sides on

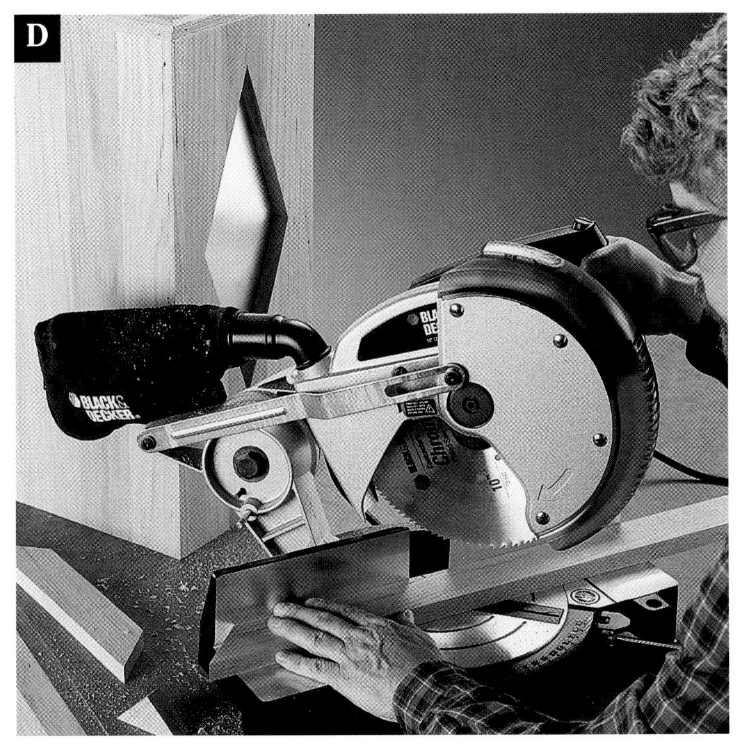


*Use a miter saw to cut 45° angles on the top trim.*

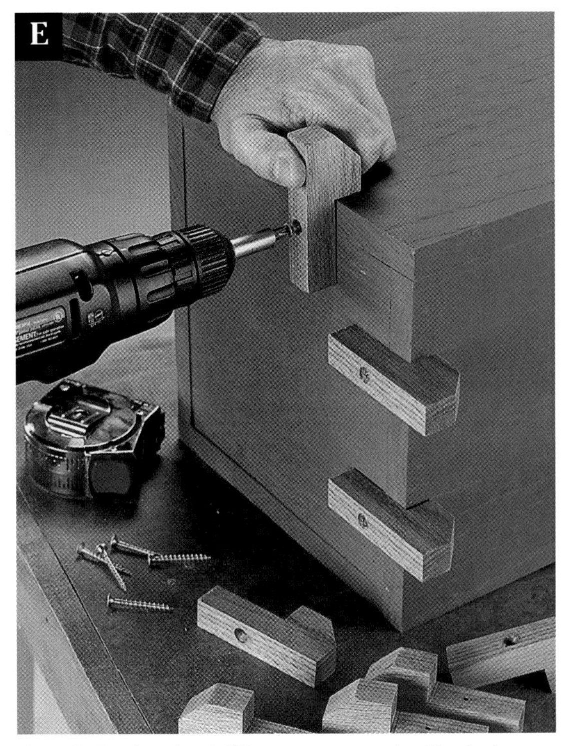


*Attach the feet by drilling countersunk pilot holes and driving 1¼" wood screws.*

your worksurface, then butt one of the cutout pieces upright at a 90° angle against the left edge (make sure the tin panel is not covered). Drill countersunk pilot holes, then join the pieces together with wood glue and 1¼" screws. Rotate the assembly so the side with the cutout is facedown, then butt the other plain side against the left edge, and attach with glue and screws driven into countersunk pilot holes. Position the bottom piece inside the assembly, flush with the bottom edges. Attach with glue and wood screws driven through countersunk pilot holes. Rotate the assembly and attach the other cutout side with glue and screws.

ATTACH THE COVE MOLDING AND TOP TRIM. The cove molding is mitered on the ends and lock-nailed together to prevent separation (see *Tip*).

Cut the cove molding (E) to size, mitering the ends at 45° angles. Position the molding so the top edges are flush with the tops of the sides, then drill pilot holes and attach the molding with glue and 1" brads **(photo C).** "Lock-nail" the cove joints with brads.

Cut the top trim (D) to size, and miter the ends at 45° **(photo D).** Position the trim so it overhangs the outer edges of the cove by ¼", then drill pilot holes and attach with glue and 4d finish nails. Lock-nail the mitered ends with 2d finish nails. Recess all the nail heads with a nail set.

CUT THE FEET. Cut the blanks for the feet, then use a jig saw to trim off the corners and make the notches (see *Diagram*). Sand the cut edges smooth.

APPLY FINISHING TOUCHES. Fill the nail and screw holes with putty. Sand the wood, then finish as desired.

To paint the sides, mask off the cove and trim and cover the tin panel with contact paper. When the paint dries, remove the contact paper and apply an amber shellac to "age" the tin. Mask off the sides, then apply finish to the cove, trim and feet. When the finish is dry, attach the feet with wood screws driven through countersunk pilot holes **(photo E).**

### TIP

*Lock-nailing is a technique used to reinforce mitered joints. The idea is to drive finish nails through both mating surfaces at the joint. Start by drilling pilot holes all the way through one board (to avoid splitting the wood) and partway into the other mating surface. Drive a small finish nail (2d or 4d) through each pilot hole to complete the lock-nailing operation.*

# Kids' Coat Rack

*Kids love using this monkey-topped coat rack, and you'll have fun building it.*

PROJECT POWER TOOLS

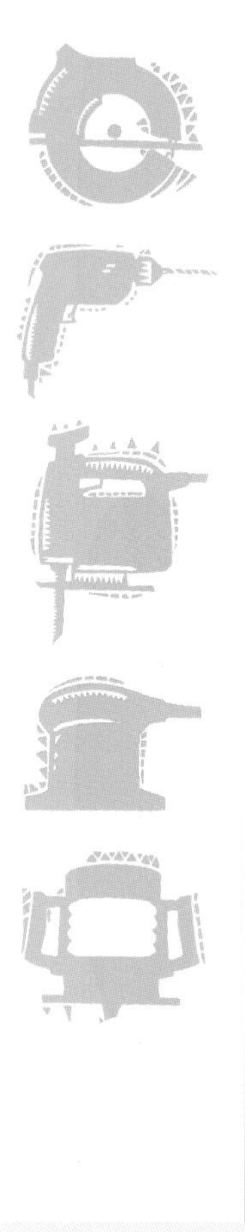

CONSTRUCTION MATERIALS

| Quantity | Lumber |
|---|---|
| 1 | 2 × 2" × 4' oak |
| 1 | 1 × 6" × 2' oak |
| 1 | ¾" × 2 × 2' birch plywood |
| 1 | 1 × 3" × 4' oak |

This stand is designed to hold eight coats or jackets, but kids will hang almost anything on the shaker pegs, including mittens, scarves, sweaters and pants. The decorative monkey acts as a motivator and reminder that it's more fun to hang your clothes on the stand than throw them on the furniture or floor. The monkey also gives you an opportunity to put your artistic talents to work. This popular stand is easy to construct and takes up little space, so it can fit in an entryway or bedroom with ease.

OVERALL SIZE:
58½" HIGH
16" WIDE
16" DEEP

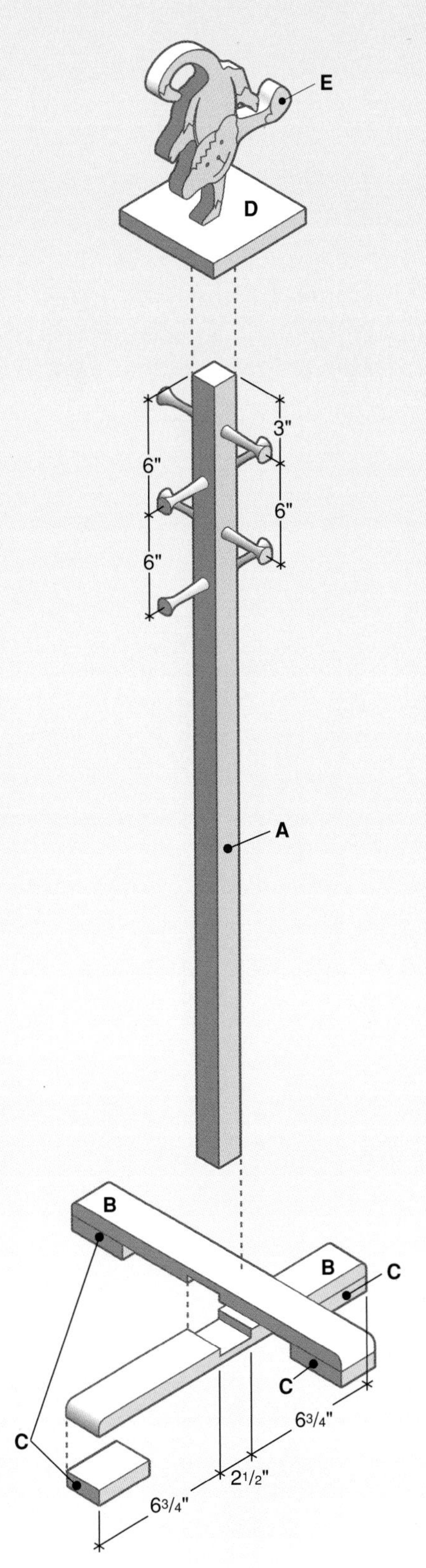


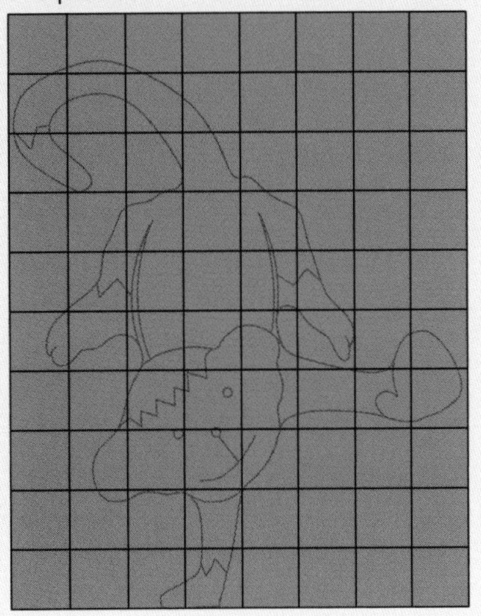


**CUTOUT DETAIL**

**Cutting List**

| Key | Part | Dimension | Pcs. | Material |
|---|---|---|---|---|
| A | Post | 1½ × 1½ × 46¼" | 1 | Oak |
| B | Leg | ¾ × 2½ × 16" | 2 | Oak |
| C | Foot | ¾ × 2½ × 3" | 4 | Oak |
| D | Platform | ¾ × 5½ × 5½" | 1 | Oak |
| E | Monkey | ¾ × 8 × 10" | 1 | Birch plywood |

**Materials:** Wood glue, #8 × 1¼" wood screws, birch shaker pegs (8), finishing materials.

**Note:** Measurements reflect the actual thickness of dimensional lumber.

*Clean out the half lap joints with a chisel and a hammer. To ensure a tight fit, make sure to keep the edges square.*

*Mark and drill holes in the post to match the diameter of your shaker pegs.*

**TIP**

*To ensure accurate cuts, build a shooting board from a straight piece of 1 × 4" lumber about 24" long, and a smooth piece of ¼"-thick plywood about 6" wide and 24" long. Attach the 1 × 4" board along one edge of the plywood strip, using glue and screws. Then, run your circular saw along the 1 × 4" straightedge, trimming the plywood base to the exact distance between the edge of the saw foot and the blade. To use the shooting board, simply clamp it in place with the edge of the plywood along the cutting line, then run your saw over the plywood with the base of the saw tight against the straightedge.*

## Directions: Kids' Coat Rack

CUT THE COAT RACK PARTS. Cut the post (A), legs (B) and feet (C) to length. Align the legs side by side, and clamp together. Mark a 2½"-wide notch on each leg (see *Diagram*). Build a shooting board (see *Tip*), and set the depth of the saw blade at a depth of ⅝" (allowing for the ¼"-thick plywood base, this will give you a ⅜"-deep cut). Clamp the shooting board next to one side of the notch and make the first cut, keeping the saw base flat on the plywood and tight against the straightedge. Reposition the shooting board and cut the other side of the notch. Leave the shooting board in place after the second cut, and make additional cuts within the notch to remove the wood between the first two cuts. Carefully clean any waste from the notch with a sharp ¾" chisel **(photo A).**

Test-fit the legs. If necessary, adjust the lap joint by chiseling, filing or sanding more stock from the notches. Round off the top edges of the leg ends with a router or belt sander.

ASSEMBLE THE PARTS. Glue and clamp the feet to the legs. Position the post on the leg assembly by drawing intersecting

*Drill a pilot hole into the base of the monkey so it doesn't split when attaching it to the platform.*

*Attach the monkey to the platform using glue and a wood screw.*

diagonal lines across the notch, then aligning each corner of the post on one of these lines. Drill two countersunk pilot holes through the bottom of the leg assembly, then attach the legs to the post with glue and wood screws.

Mark two peg holes on each side of the post (see *Diagram*). Carefully drill holes straight into the sides of the post, matching the diameter of the shaker pegs **(photo B).**

MAKE THE MONKEY AND PLATFORM. Lay out the monkey pattern (E) on birch plywood (see *Diagram*), and cut out the pattern with a jig saw. Use wood putty to fill any voids on the edges of the plywood. Cut the platform (D) to size.

Drill a countersunk pilot hole into the bottom of the platform at the centerpoint for attaching the monkey. Drill two offset pilot holes in the top of the platform, about ¾" from the center hole. Counterbore one of these holes. Drill a pilot hole into the center of the monkey's paw **(photo C).** Paint the monkey with brown and white paint.

ASSEMBLE THE UNIT. First attach the monkey to the platform, using glue and a wood screw **(photo D).** Then, attach the monkey and platform to the post assembly with glue, a wood screw and a brad. Attach the shaker pegs with glue, and wipe off any excess.

APPLY THE FINISHING TOUCHES. Sand the project smooth and apply oil or a clear finish.

TIP

*For better control when painting faces and figurines, use latex paint as a base coat, and outline the pattern details with a permanent-ink marker. To protect your work, you can seal the monkey with a low-luster water-based polyurethane.*

PROJECT
POWER TOOLS

# Dry Sink

*This classic cabinet brings antique charm to any setting.*

CONSTRUCTION MATERIALS

| Quantity | Lumber |
|---|---|
| 4 | 1 × 2" × 6' birch |
| 5 | 1 × 3" × 8' birch |
| 1 | 1 × 4" × 8' birch |
| 2 | 1 × 6" × 6' birch |
| 1 | ½" × 4 × 4' birch plywood |
| 1 | ¾" × 4 × 8' birch plywood |
| 3 | ⅜" × 4' birch dowling |

A traditional dry sink was used to hold a washbasin in the days before indoor plumbing, but today it can serve a variety of decorative and practical functions around the house. Our classic dry sink is used as a garden potting table. It's the ideal height for mixing soils, planting seeds and watering plants. The top has a handy back shelf to hold plants and accessories, while the curved front and sides are especially designed to contain messy spills. The roomy cabinet has two hinged doors for easy access and enough interior space to store pots, planters, fertilizers, insecticides and an assortment of gardening tools. This project features birch plywood panels with solid birch frames secured with strong "through-dowel" joinery.

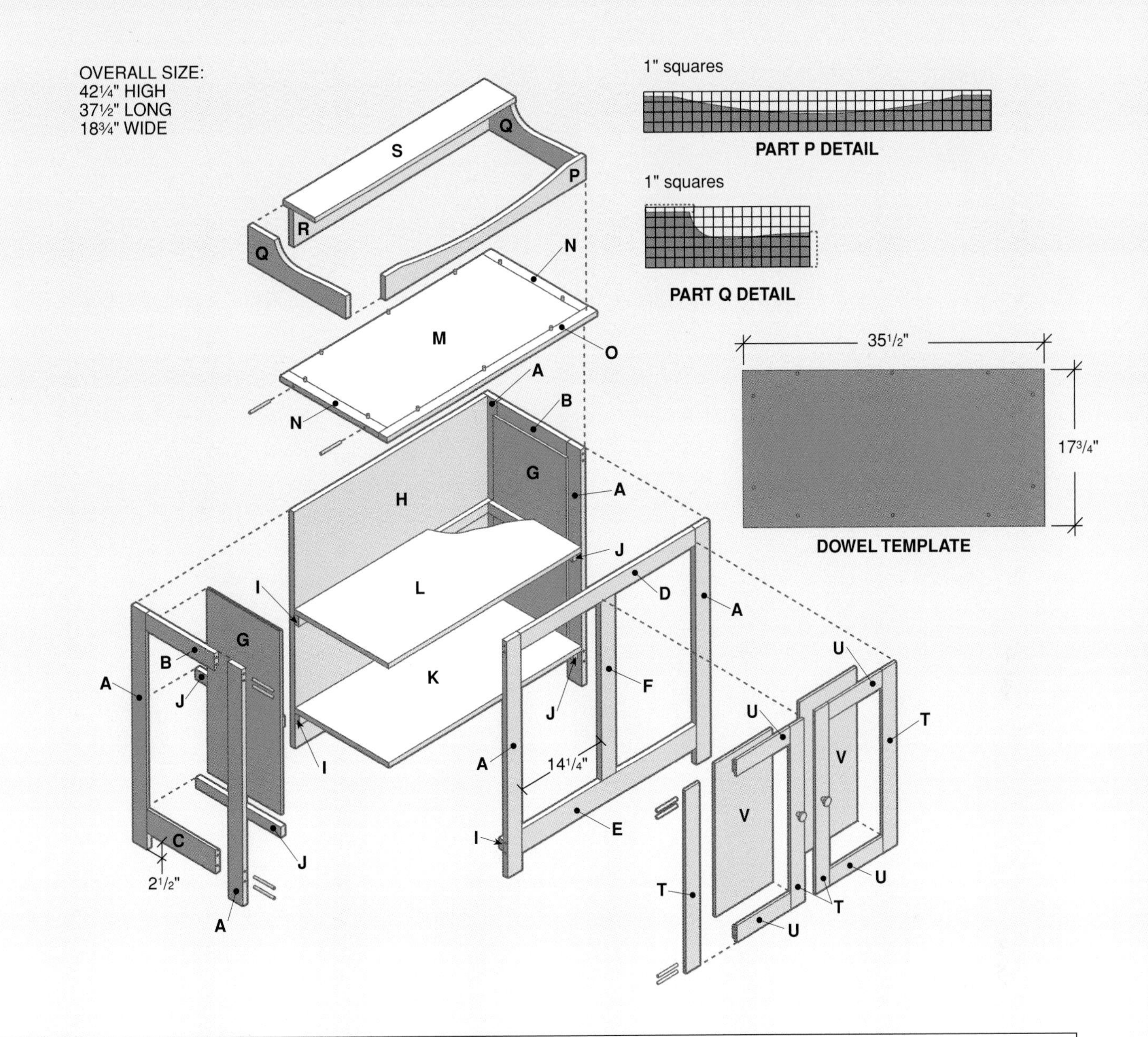


| Key | Part | Dimension | Pcs. | Material |
|---|---|---|---|---|
| A | Stile | ¾ × 2½ × 35¼" | 6 | Birch |
| B | Side rail, top | ¾ × 2½ × 12¼" | 2 | Birch |
| C | Side rail, bottom | ¾ × 3½ × 12¼" | 2 | Birch |
| D | Front rail, top | ¾ × 2½ × 31" | 1 | Birch |
| E | Front rail, bottom | ¾ × 3½ × 31" | 1 | Birch |
| F | Mullion | ¾ × 2½ × 26¾" | 1 | Birch |
| G | Side panel | ½ × 12⅞ × 27½" | 2 | Birch ply. |
| H | Back panel | ¾ × 34½ × 35¼" | 1 | Birch ply. |
| I | Back, front cleat | ¾ × 1½ × 34½" | 3 | Birch |
| J | Side cleat | ¾ × 1½ × 15" | 4 | Birch |
| K | Bottom | ¾ × 16½ × 34⅜" | 1 | Birch ply. |

| Key | Part | Dimension | Pcs. | Material |
|---|---|---|---|---|
| L | Shelf | ¾ × 16⅜ × 34⅜" | 1 | Birch ply. |
| M | Top | ¾ × 17¼ × 34½" | 1 | Birch ply. |
| N | Top side edge | ¾ × 1½ × 17¼" | 2 | Birch |
| O | Top front edge | ¾ × 1½ × 37½" | 1 | Birch |
| P | Top assem. front | ¾ × 3½ × 35¾" | 1 | Birch |
| Q | Top assem. side | ¾ × 5½ × 17" | 2 | Birch |
| R | Top assem. back | ¾ × 5½ × 34" | 1 | Birch |
| S | Top assem. cap | ¾ × 5½ × 35¾" | 1 | Birch |
| T | Door stile | ¾ × 2½ × 27¼" | 4 | Birch |
| U | Door rail | ¾ × 2½ × 9¾" | 4 | Birch |
| V | Door panel | ½ × 10½ × 23" | 2 | Birch ply. |

**Materials:** Wood glue, birch shelf nosing (⅛ × ¾ × 34½"), 16-ga. brads, #8 wood screws (1¼", 1⅝"), 4d finish nails, ⅜" inset hinges (4), ⅜ × 1" dowels (10), ⅜ × 3" dowels (40), door pulls (2), finishing materials.

**Note:** Measurements reflect the actual thickness of dimensional lumber.

*Drill holes and insert the dowels after the frame pieces have been glued together.*

*Carefully square the corners of the rabbets in the side panels and door panels, using a sharp chisel.*

TIP

*Custom-cut dowels can be difficult to fit into snug holes. To make insertion easier, score a groove along the length of each dowel. This groove allows air and excess glue to escape when driving the dowels into their holes.*

## Directions: Dry Sink

CUT AND ASSEMBLE THE CABINET FRAMES. The dry sink is built with two side frames, a face frame and two door frames—all made from birch rails and stiles joined with "through" dowels.

Begin by cutting the stiles (A) and rails (B, C, D, E) and mullion (F) to length.

Build each side frame by gluing a top and bottom side rail between two stiles. The bottom rail should be raised 2½" from the bottoms of the stiles. Clamp in place, check for square and let dry.

After the glue has dried, drill two ⅜"-dia. × 3"-deep holes through the stiles at each rail location **(photo A).** Cut 3"-long dowels, and score a groove along one side. Apply glue to the dowels, then use a mallet to drive them into the holes.

Repeat this process to construct the front frame, using two stiles, the top and bottom front rails, and the mullion. Make sure the mullion is centered between the stiles.

Cut the door frame stiles (T) and rails (U) to size with a circular saw, and assemble frame parts in a similar fashion.

*Drill pilot holes, and attach cleats with wood screws to the frame but not into the ½" panel.*

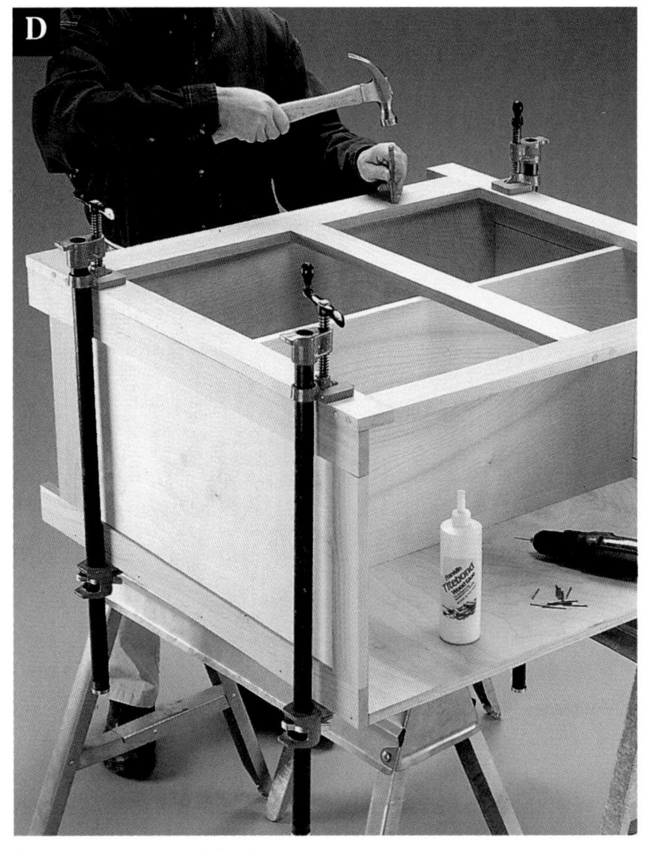


*Clamp one side of the face frame and check for square, then clamp the other side and check for square again. Attach with finish nails and set.*

ADD THE PANELS. The ½" plywood side panels and door panels fit into rabbets cut around the inside of the side frames and door frames.

Mount a ⅜" rabbet bit in your router, set to ½" depth. Cut a continuous rabbet around the inside of the side frames. Square off the corners of the rabbet, using a chisel **(photo B).** On the back face of each door frame, cut a rabbet around the inside of the frame in a similar fashion.

Next, change the depth of the router bit to ⅜", and cut another rabbet around the outside edge of the door frame. This creates a lip which will overlap the face frame when the doors are attached.

Cut the side panels (G) and the door panels (V) to size. Position each panel inside its frame, then drill pilot holes and attach the panels with 16-ga. brads. Position and attach hinges and knobs on the cabinet doors.

PREPARE THE REMAINING PIECES. Cut the back panel (H), bottom (K) and shelf (L) to size. Cut and attach shelf nosing to the front edge of the shelf, using glue and brads.

Cut the front and back cleats (I) and side cleats (J) to size. On the inside faces of the face frame stiles, mark reference lines 5¼" from the bottom. On the inside faces of the back panel and side frame stiles, mark reference lines at 5¼" and 21" from the bottom.

To attach the side cleats, position the cleats with the top edges flush with the reference lines, with the ends of cleats set back ¾" from the front edge and 1½" from the back edge. Drill countersunk pilot holes, and attach the cleats with 1¼" wood screws **(photo C).** NOTE: Take care to screw the side cleats into the frame members only, not into the ½" panels.

Attach the back cleats to the back panel, and the front cleat to the inside of the face frame,

*Cut the top side edges and top front edge, and attach the pieces with wood glue. Clamp in place until the glue dries, then drill and insert dowels to strengthen the joints.*

*When creating the top assembly, first attach the front piece to the sides, then attach the sides to the back, using wood glue and 4d finish nails.*

using the same process.

ASSEMBLE THE CABINET. Position the back panel between the side assemblies. Drill countersunk pilot holes, and attach the sides to the back panel with 1⅝" wood screws. Next, position the bottom over the cleats. Check to make sure the cabinet is square, then drill pilot holes and attach the bottom by driving 4d finish nails into the cleats. Position and attach the shelf in the same manner.

Lay the cabinet on its back and clamp the face frame in position. Check for square, then drill pilot holes and attach the face frame to the cabinet with glue and 4d finish nails driven into the side frames, bottom and shelf **(photo D).** Also drive finish nails through the bottom and into the front cleat. Set all nail heads.

Position and mount the doors in their openings, then remove them and detach the hinges and knobs until the wood has been finished.

ASSEMBLE AND ATTACH THE TOP. Cut cabinet top (M), side edges (N) and front edge (O) to size. Attach the edges around the top, using glue. Clamp the pieces in place until the glue dries **(photo E).** After the glue has dried, drill holes and reinforce the joints with 3"-long dowels, following the same procedure used to construct the cabinet frames.

Position the top on the cabinet, leaving a ¾" overhang on both ends and the front. Drill countersunk pilot holes, and attach the top with glue and 1⅝" wood screws driven into the cabinet frames and back panel.

CREATE THE TOP ASSEMBLY. Cut the top assembly parts (P, Q, R and S) to size. Transfer the patterns to the pieces (see *Diagram*), then cut them out with a jig saw. Sand the cut edges smooth.

Position the front piece against the side pieces, so there is a ⅛" overhang on both ends. Drill pilot holes and attach the front piece to the side pieces with glue and 4d finish nails.

*Use a template to ensure that the dowel holes in the top assembly will match those drilled in the top of the cabinet.*

Position the back piece between the sides, then drill pilot holes and attach the pieces with glue and 4d finish nails **(photo F).** Position the cap piece so it overlaps by ⅛" on each end, then drill pilot holes, and attach with glue and 4d finish nails.

ATTACH THE TOP ASSEMBLY. The top assembly is attached to the cabinet with dowels, positioned with the benefit of an easy-to-build template.

Create the template by tracing the outer outline of the top assembly on a piece of scrap plywood or hardboard. Cut the template to this size. Place the template over the bottom of the top assembly, then drill ⅜"-dia. × ½"-deep dowel holes through the template and into the top assembly.

Place the template on the cabinet top, centered side to side with the back edges flush, and tape it in place. Drill corresponding dowel holes into the top **(photo G).** Remove the template and attach the top assembly to the cabinet with glue and 1" dowels. Use weights or clamps to hold the top assembly in place until the glue dries.

FINISH THE CABINET. To give our dry sink a vintage look, we used a unique antiquing method that uses both stain and paint.

First, finish-sand all surfaces and edges of the cabinet. Next, stain the entire project (we used medium cherry stain). After the stain dries, apply latex paint to the desired surfaces (we used Glidden® Centurian blue paint, applying it to all surfaces except the top).

While the paint is still damp, use a cloth and denatured alcohol to remove color until you achieve the desired look. To mimic the look of a genuine antique, try to remove most of the paint from the corners and edges, where a cabinet typically receives the most wear. Reinstall the hardware and hang the doors after the finish has dried.

**TIP**

*Custom finishes can be tricky to create. Before attempting a unique finish, like the stain/rubbed-paint treatment used on our project, always practice on scrap materials.*

PROJECT
POWER TOOLS

# Candy Dish

*Attractive contrasting woods give this candy dish its delicious appearance.*

### Construction Materials

| Quantity | Lumber |
|---|---|
| 1 | ½ × ½" × 4' birch |
| 1 | ½ × 3¾" × 4' birch |
| 1 | ½ × 12 × 30" birch plywood |
| 1 | ¼ × 1¾" × 4' walnut |
| 1 | ½ × 1¾ × 30" walnut |
| 1 | 1"-dia. × 36" walnut dowel |

This distinctive candy dish features a complementary combination of woods: birch and walnut. The naturally dark walnut side trim creates a pleasing contrast with the lighter birch box. The removable walnut divider also lets you create compartments for separating several kinds of snacks. Or, if you prefer, you can use the dish without the divider to display a single treat. Striking, yet easy to build, this candy dish makes such a great gift that you may want to capitalize on your set-up time and build several at once.

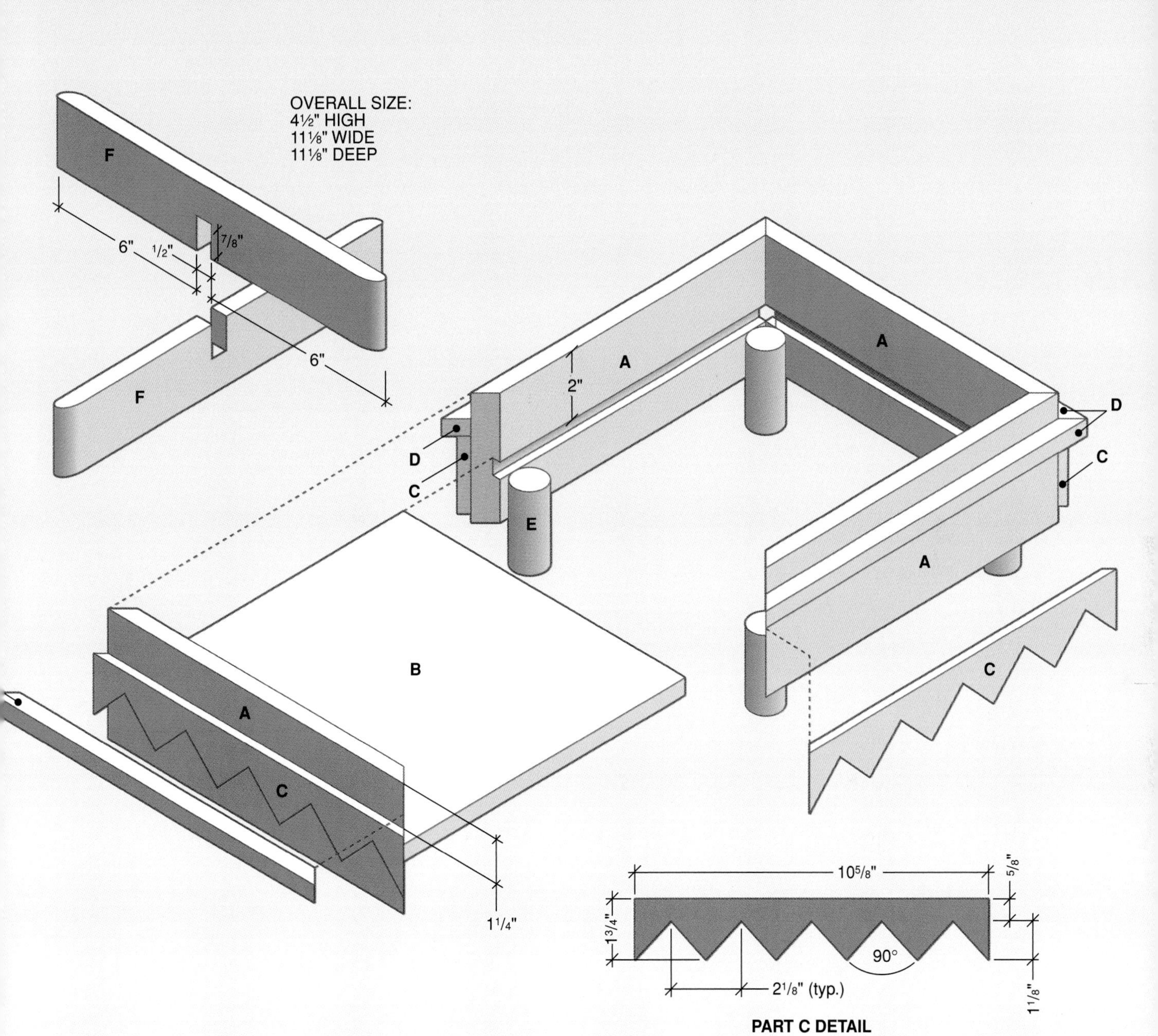


**Cutting List**

| Key | Part | Dimension | Pcs. | Material |
|---|---|---|---|---|
| A | Box side | ½ × 3¾ × 10⅛" | 4 | Birch |
| B | Bottom | ½ × 9⅝ × 9⅝" | 1 | Birch plywood |
| C | Side trim | ¼ × 1¾ × 10⅝" | 4 | Walnut |
| D | Side trim cap | ½ × ½ × 11⅛" | 4 | Birch |
| E | Foot | 1"-dia. × 2" | 4 | Walnut dowel |
| F | Divider | ½ × 1¾ × 12½" | 2 | Walnut |

**Materials:** Wood glue, 16-ga. brads, finishing materials.

**Note:** Measurements reflect the actual thickness of dimensional lumber.

*Use bar clamps and a straightedge to hold pieces in place and ensure uniform dadoes.*

*Assembly is easier if you first glue and band-clamp the box sides, and then nail the sides together.*

## Directions: Candy Dish

CREATE THE BOX SIDES AND BOTTOM. Dadoes are cut in the box sides to secure the bottom.

Cut the box sides (A) to length, mitering ends at 45°. Then, cut the dadoes, using a straight-line routing jig to make uniform, straight cuts. To make the jig, clamp two pieces of ½"-thick scrap wood against the edges of a box side. Securely clamp a straightedge over the scrap wood to guide the router base and cut the dado at the correct location (see *Diagram*) with a ½" straight-cutting router bit set ¼" deep (**photo A**). Repeat this process (see *Tip*) to cut dadoes in all four sides. Be sure to cut all dadoes on the inside faces.

ASSEMBLE THE BOX. Cut the bottom (B) to size, and test-fit the sides and bottom. Apply glue to the miters, then clamp the sides in position around the bottom. Drill pilot holes, and secure the pieces with brads driven through the miters **(photo B).**

CUT AND ASSEMBLE THE SIDE TRIM AND CAP. On the 48" length of ¼" walnut stock, draw a reference line ⅝" down from the top edge along the entire length of the piece. Cut the side trim pieces (C) to length, miter the corners at 45°, and test-fit. Create the decorative sawtooth pattern (see *Diagram*) on one side trim piece with a combination square, working from both ends toward the middle. The pattern lines should be at 45° angles to the bottom edge, and the top corner of the pattern should touch the reference line. Cut out the notches with a jig saw. Using this piece as a template, transfer the pattern to the remaining side trim blanks. Cut the remaining notches **(photo C).** Position the trim pieces on the sides (see *Diagram*), drill pilot holes, and fasten to the box with glue and brads. Cut the side trim caps (D), miter the ends, and secure with glue.

Cut the feet (E) from walnut dowels, and sand smooth. Attach the feet with glue applied liberally at all three points where dowel contacts box **(photo D).**

> TIP
>
> *Once you have set up a straight-line routing jig, take full advantage of it by machining many identical pieces efficiently. To shape a new piece, simply loosen the bar clamps, insert a new blank, and reclamp; no need to reposition the straightedge.*

*Securely clamp the side trim and cut several notches, then rotate the piece, reclamp, and cut the remaining notches.*

*Attach the feet, using glue at each box contact point.*

*Check the divider for square, using a triangle, and test-fit into the box, making sure you can easily remove it for cleaning.*

CUT AND ASSEMBLE THE INTERLOCKING DIVIDERS. Cut the dividers (F) to length and cut a ½"-wide × ⅞"-deep notch at the center of each divider, using a jig saw. Round over the divider ends with a belt sander, and interlock the two divider pieces. Check for square **(photo E),** and test-fit the divider to make sure it fits in the box. Sand the pieces smooth. This will make it easier to remove the divider unit for cleaning.

APPLY FINISHING TOUCHES. Finish the divider and box with clear, water-based polyurethane.

TIP

*To prevent the pieces from "welding" together and making cleaning more difficult, finish the dish and divider pieces separately. Allow plenty of time for the finish to cure before putting them together.*

PROJECT
POWER TOOLS

# Two-tier Bookshelf

*Here's a smart-looking, easy-to-build project that has no glue, screws or nails!*

CONSTRUCTION MATERIALS

| Quantity | Lumber |
|---|---|
| 1 | ¾" × 4 × 8' Baltic birch plywood |
| 1 | 1"-dia. × 2' birch dowel |

This two-tier bookshelf provides ample room for encyclopedias, dictionaries and other useful references. The modern side profile complements many decorating motifs, and with the right finish, this project can become a vibrant accent piece. The bookshelf uses an unusual joinery method, known as a *pinned mortise-and-tenon,* that requires no glue, screws or nails in the assembly. Instead, wedges hold the joints together. When moving or storing the unit, you can simply remove the wedges.

With the included plan for a mortising jig, you can easily make several of these bookshelves to give as gifts.

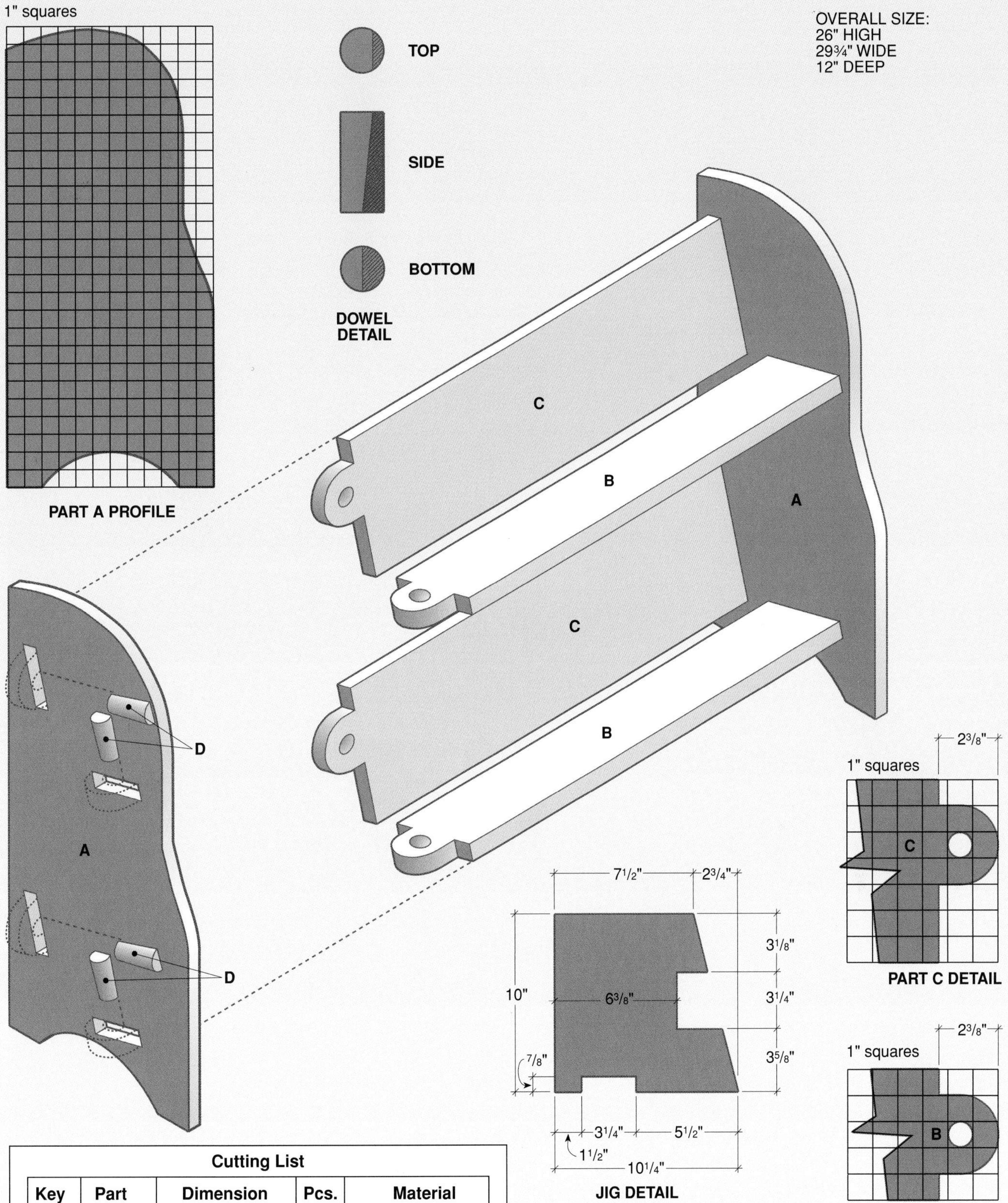


**Cutting List**

| Key | Part | Dimension | Pcs. | Material |
|---|---|---|---|---|
| **A** | End | ¾ × 12 × 26" | 2 | Birch plywood |
| **B** | Shelf | ¾ × 5 × 29¾" | 2 | Birch plywood |
| **C** | Back | ¾ × 7 × 29¾" | 2 | Birch plywood |
| **D** | Wedge | 1"-dia. × 2¼" | 8 | Birch dowel |

**Materials:** Finishing materials.

**Note:** Measurements reflect the actual thickness of dimensional lumber.

## Directions: Two-tier Bookshelf

MAKE THE JIG. This project uses a jig to help you accurately mark the location of mortises in the side pieces.

First, cut a 10 × 10¼" blank from ¼" scrap material. Measure and mark the diagonal line and the locations for the mortise guides (see *Diagram*). Use a jig saw to cut out the jig **(photo A).**

CUT THE ENDS. Cut the end blanks (A) to size, then transfer the pattern (see *Diagram*) and cut with a jig saw. Lay both ends on your workbench with the back edges together, forming a mirror image. Measure from the bottom back corners and mark reference points at ⅞" and 14¾" **(photo B).**

Lay out the mortises by positioning the bottom back corner of the jig at the first reference point, keeping the back edges flush. Outline the two lower mortises, then slide the jig up to the second reference point and mark the two higher mortises **(photo C).**

TIP

*To yield an opening large enough to accommodate the tenons, make sure the mortises on your pattern are slightly oversized.*

Remove the jig and draw lines to close the ¾ × 3" rectangles. Drill pilot holes and cut out the mortises with your jig saw **(photo D).**

CUT THE SHELVES AND BACKS. Cut shelves (B) and backs (C) to size. Lay out the profile for the tenons (see *Diagram*) and cut out with a jig saw. Sand the edges smooth. Drill wedge holes, using a 1" spade bit. Test-fit the tenons in the mortises, and make any necessary adjustments.

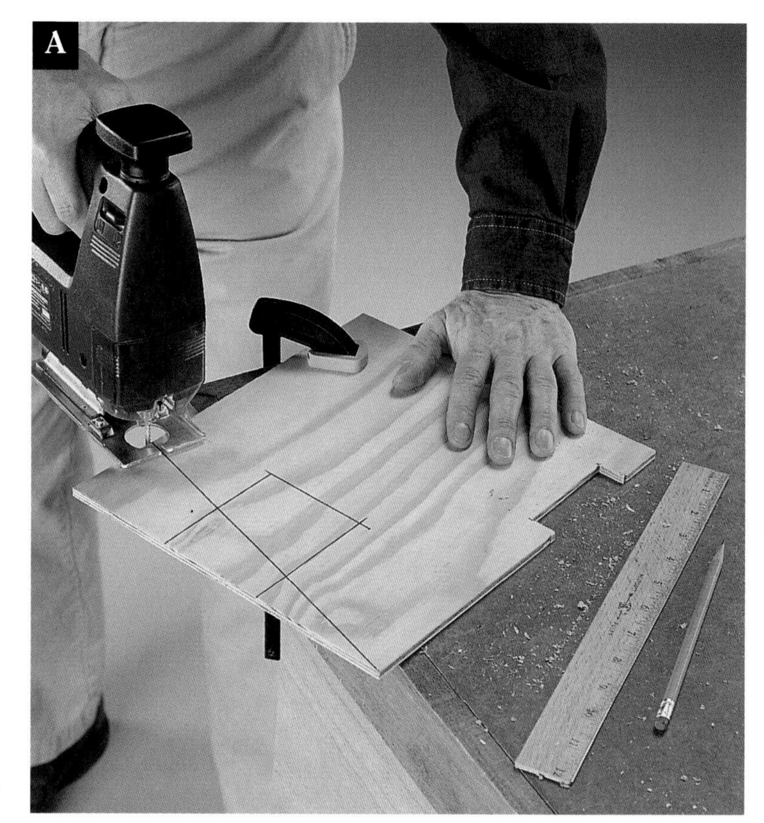

*Use a piece of ¼" scrap plywood and your jig saw to create the mortising jig.*

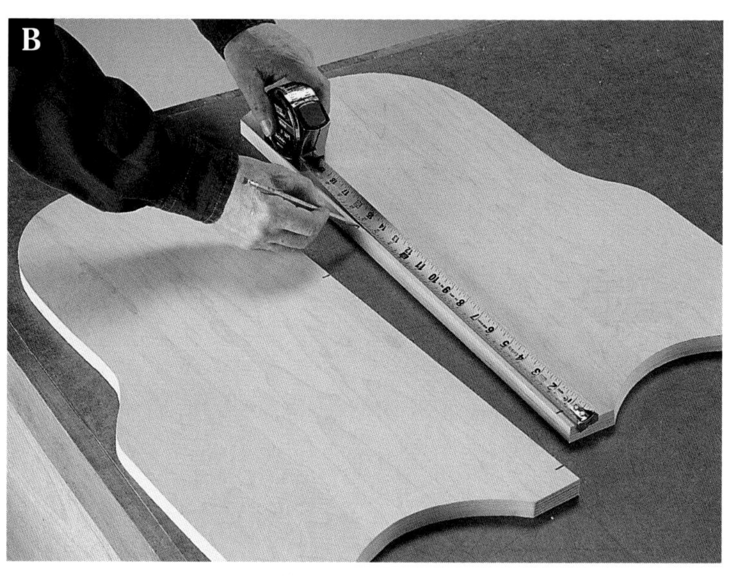

*Mark reference points at ⅞" and at 14¾" along the backs of both sides as a guide for positioning the mortise jig.*

MAKE THE WEDGES. Create wedges (D) to hold the shelves in place by cutting 1"-dia. dowels to 2¼" lengths. Measure from one edge and mark reference lines across the top of the dowel at ¼" and across the bottom at ½". Connect these lines, then sand the dowels down to this line, using a belt sander clamped horizontally to your worksurface **(photo E).**

*Use the jig to mark the locations of both pairs of mortises, then flip the jig and mark corresponding mortises on the other end piece.*

*Drill pilot holes and complete the mortises with your jig saw.*

**TIP**

*When a project has modern lines, such as this two-tier bookshelf, you can create a very interesting accent piece by applying an exotic colored aniline dye. For instance, a rich burgundy or a bright green would create a vibrant accent.*

Assemble the shelves and backs between the ends and test-fit the wedges. Disassemble the bookshelf for finishing.

APPLY THE FINISHING TOUCHES. Finish-sand the entire project, then paint or finish the bookshelf as desired. We used a light oil finish, but this project would also be well suited for a bold aniline dye. When the finish dries, assemble the pieces.

*Sand the wedges to the lines, using a belt sander clamped to your worksurface.*

PROJECT
POWER TOOLS

# Mantel

***Deceptively simple to build, this elegant mantel mimics the look of hardwood at a fraction of the cost.***

### Construction Materials

| Quantity | Lumber |
|---|---|
| 1 | 1 × 8" × 6' poplar |
| 1 | 2 × 4" × 4' poplar |
| 1 | 2 × 2" × 6' poplar |
| 1 | ¾ × 3¾" × 5' crown molding* |
| 1 | ½ × ⅝" × 5' dentil molding |

*Specialty crown molding from Aboxablox, style #621.

This mantel will receive high praise from friends and relatives. With its wide shelf and 4' length, the mantel is a great place to display family photographs or prized possessions. It is also an excellent starting point for holiday or seasonal decorating.

Though the mantel appears to be a solid piece of milled hardwood, its looks are deceiving. Stock moldings, miter cuts and lock-nailing hide a simple support framework, and an antique white paint finish disguises the mantel's use of inexpensive poplar lumber.

And don't skip this project if you don't have a fireplace. This mantel makes a wonderful display shelf anywhere in your home.

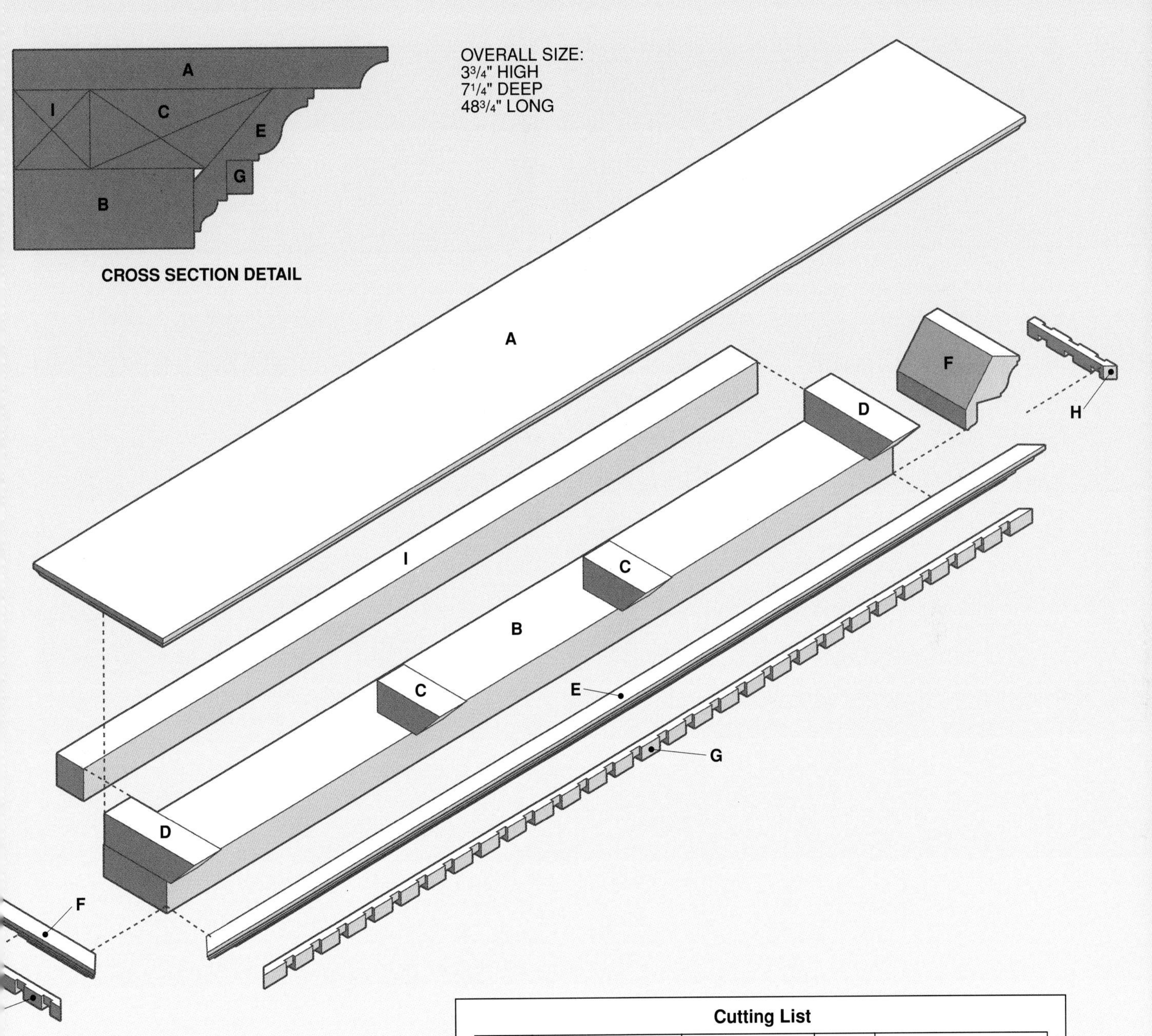


| Cutting List | | | | |
|---|---|---|---|---|
| **Key** | **Part** | **Dimension** | **Pcs.** | **Material** |
| **A** | Top | ¾ × 7¼ × 48¾" | 1 | Poplar |
| **B** | Bottom | 1½ × 3½ × 41" | 1 | Poplar |
| **C** | Center support | 1½ × 1½ × 3½" | 2 | Poplar |
| **D** | End support | 1½ × 1½ × 5" | 2 | Poplar |
| **E** | Front crown | ¾ × 3¾ × 46¼" | 1 | Crown molding |
| **F** | Side crown | ¾ × 3¾ × 6" | 2 | Crown molding |
| **G** | Front dentil | ½ × ⅝ × 43½" | 1 | Dentil molding |
| **H** | Side dentil | ½ × ⅝ × 4⅝" | 2 | Dentil molding |
| **I** | Ledger | 1½ × 1½ × 39" | 1 | Poplar |

**Materials:** Wood glue, wood screws (#8 × 2¼", #6 × 1½"), finish nails (2d, 4d), ⅜" wood plugs, finishing materials.

**Note:** Measurements reflect the actual thickness of dimensional lumber.

*Use a combination square to mark the location of the center supports on the ledger, then attach with glue and screws.*

*Attach the front crown and side crown to the bottom and supports with glue and 4d finish nails.*

## Directions: Mantel

CUT AND ASSEMBLE THE BOTTOM AND SUPPORTS. Cut the bottom (B), center supports (C) and end supports (D) to size. Miter one end of each support at 45°. On the bottom, make marks 13" and 14½" in from each end, and use your combination square to draw a line at each mark. Position the end supports (see *Diagram*), drill countersunk pilot holes, and attach with glue and 2¼" wood screws.

Position the center supports along the reference lines, drill countersunk pilot holes, and fasten with glue and 2¼" screws **(photo A).** Cut the ledger (I) to length, and test-fit it between the end supports so the back edges of the ledger and bottom are flush.

ATTACH THE CROWN MOLDING. Once the bottom and the supports have been assembled, the front and side crown molding can be attached.

Cut the front crown (E) and side crown (F) to size. Miter the ends at 45° by positioning the molding upside down in a power miter box, with one flat lip against the base of the saw, and the other lip against the saw fence.

Position the front crown so the top edge is flush with the top edge of the supports and the lower edge rests against the edge of the bottom. Drill pilot holes and attach the front crown to the supports and the bottom with wood glue and 4d finish nails **(photo B).** Then, attach the side crowns in similar fashion, and "lock-nail" the crown molding joints with 2d finish nails (see *Tip*, page 35). Set all the nail heads.

ATTACH THE TOP. The front and side edges of the top are rounded over with a router.

Cut the top (A) to size and sand smooth with medium-grit sandpaper. Use a router with a ⅜" roundover bit set for a ⅛" shoulder to shape the ends and front edge (first test the cut on a piece of scrap wood).

### TIP

*We chose inexpensive poplar for our primary building material. If a more natural look is desired, you can also build the mantel from oak and finish it with stain.*

*For placement of drill holes, measure and mark the location of the supports on the top.*

*Keep the mitered joints tight when attaching the dentil molding.*

Place the top facedown on your worksurface, and mark the positions of the supports on the underside **(photo C).** Drill pilot holes and attach the top to the supports with glue and 4d finish nails. Set all nail heads.

ATTACH THE DENTIL MOLDING. Cut the front dentil (G) and side dentil (H) to size, mitering the mating ends at 45° angles. Be sure to cut through the thicker "tooth" portion of each molding piece to ensure that the repeat pattern will match at the corners.

Position the front dentil molding and side dentil molding on the crown molding (see *Diagram*). Drill pilot holes and attach with glue and 4d finish nails, keeping the mitered joints tight **(photo D).** Lock-nail the joints with 2d finish nails, and set the nail heads.

MOUNT THE MANTEL. When completed, the mantel is attached to the ledger. Anchor the ledger to the wall and test-hang the mantel before finishing it.

Begin by positioning the ledger on the wall, checking for level. Drill pilot holes and attach the ledger to the wall with 3½" wallboard screws driven into studs. Fit the mantel over the ledger and drill counterbored pilot holes through the top into the ledger.

APPLY FINISHING TOUCHES. Remove the mantel from the wall and apply putty to all nail holes. Scrape off any excess glue. Finish-sand the mantel, then apply your finish. (We chose a glossy antique white paint.) When the finish is dry, position the mantel over the ledger and mount it with 1½" screws driven into the counterbored holes. Insert glued plugs into the holes, sand, and touch up the finish.

TIP

*The mantel can be shortened or lengthened to meet your needs. To resize, simply adjust the sizes of the top, bottom, moldings and ledger accordingly.*

PROJECT
POWER TOOLS

# Window Valance

*Dress up any window with this ornate window valance.*

| Construction Materials | |
|---|---|
| **Quantity** | **Lumber** |
| 1 | 1 × 6" × 8' oak |
| 1 | 1 × 2" × 3' oak |
| 1 | 1½"-dia. × 36" oak dowel |

With its detailed profile, this valance will make any window a showpiece. It adds interest to stark windows and walls, and lends itself to a wide variety of decorating motifs. The valance can be stained, painted or covered with cloth to match any decor. It also doubles as a shelf to hold plants, collectibles and other accent pieces.

As shown here, the window valance is designed to fit over a standard 30" window, but you'll find it easy to adapt the design to larger or smaller windows.

OVERALL SIZE:
7½" HIGH
6¼" WIDE
36" LONG

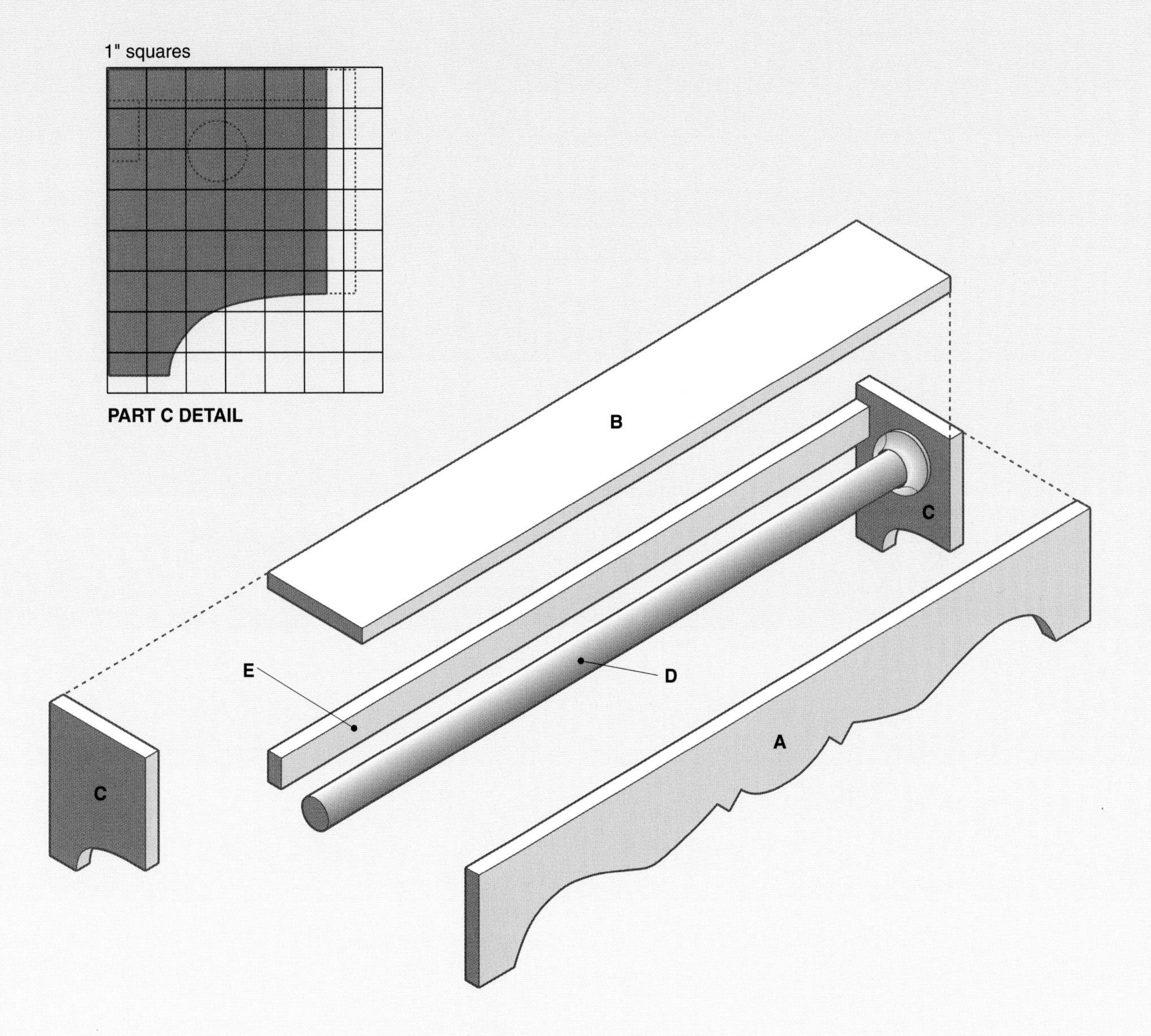


| Cutting List | | | | |
|---|---|---|---|---|
| **Key** | **Part** | **Dimension** | **Pcs.** | **Material** |
| **A** | Front | ¾ × 5½ × 36" | 1 | Oak |
| **B** | Top | ¾ × 5½ × 34½" | 1 | Oak |
| **C** | Side | ¾ × 5½ × 7½" | 2 | Oak |
| **D** | Dowel | 1½"-dia. × 34" | 1 | Oak dowel |
| **E** | Hanging strip | ¾ × 1½ × 34½" | 1 | Oak |

**Materials:** Wood glue, 4d finish nails, #6 × 2" wood screws, putty, plastic dowel rod pockets (2) with screws, 2½" wallboard screws, finishing materials.

**Note:** Measurements reflect the actual thickness of dimensional lumber.

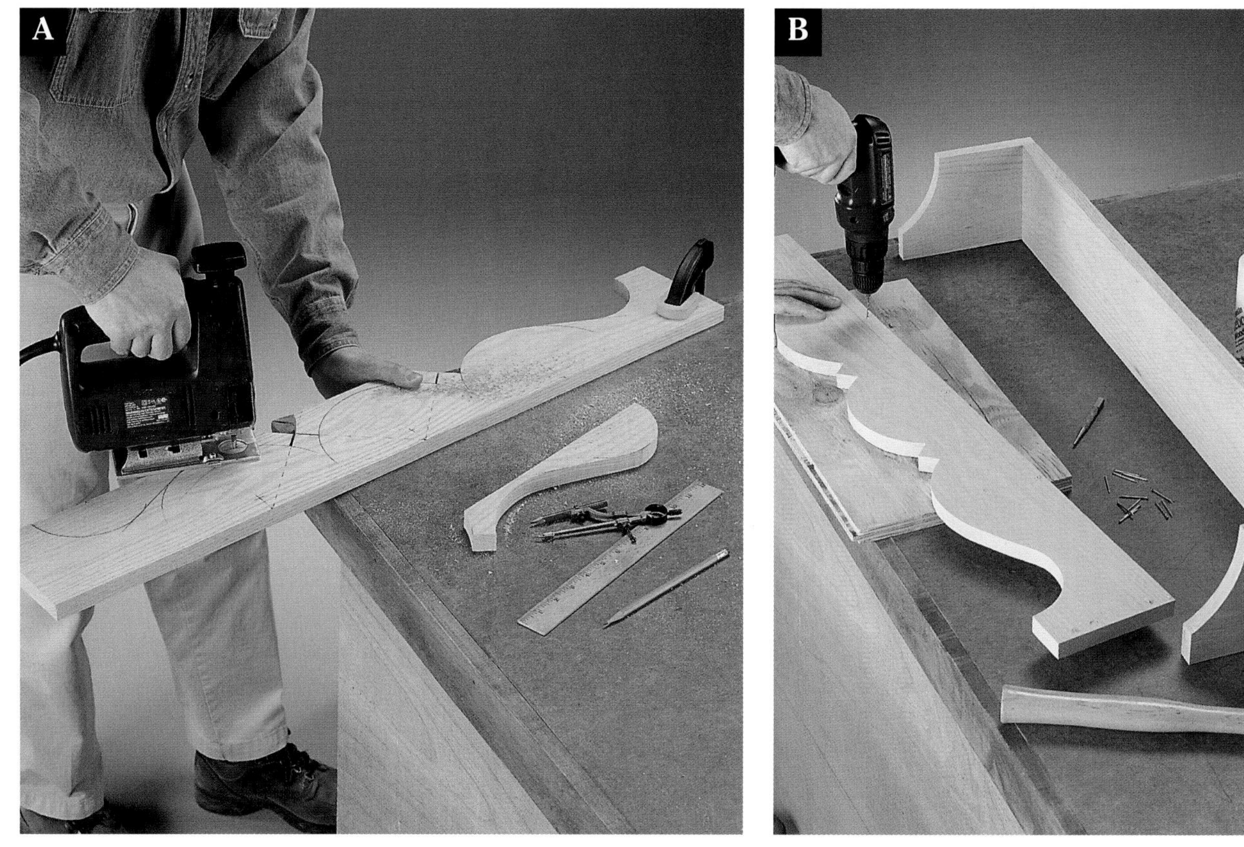

*After laying out the front pattern, clamp it firmly and cut with a jig saw.*

*Drill pilot holes in the front and sides before assembly.*

**Directions: Window Valance**

CUT AND ASSEMBLE THE PIECES. Cut the front (A), top (B), sides (C) and hanging strip (E) to size. Transfer the patterns to the front and sides (see *Diagram* and *Part A Detail*). To use the front pattern, position it on the left side of the blank, trace the outline, then flip the pattern to outline the mirrored right side of the profile. Cut the profiles with a jig saw **(photo A).** Sand all edges smooth. Position the sides, top and front together, then drill pilot holes; assemble the pieces with glue and finish nails **(photo B).**

Test-fit the hanging strip between the sides, then drill countersunk pilot holes through the top and into the hanging strip and attach the strip with 2" screws **(photo C).** (The strip will be temporarily detached when you hang the valance.)

ATTACH THE ROD. Drill pilot holes and screw the plastic dowel rod pockets to the sides **(photo D).** Position the rod pockets toward the front of the valance, so they will be hidden by the front of the valance when the unit is hung. NOTE: The position of the rod pockets may be dictated by the style of drapes you will be hanging; make sure to check on this before attaching the pockets.

Set all nail heads and fill with putty. Sand the entire project smooth. Cut the dowel (D) to length and place in the rod pockets.

APPLY FINISHING TOUCHES. It will be easiest to finish the valance before hanging it. Make sure to detach the hanging strip and rod while you finish the wood. We used a rustic oak finish with a water-based polyurethane top coating, but you can choose another finish that matches the decor of your rooms or best suits your personal taste.

HANG THE VALANCE. Detach the hanging strip from the

TIP

*For a more professional look, build your valance from the same type of wood and finish used to trim your windows. If you plan on painting the valance, use yellow poplar, which is readily available at budget lumberyards.*

*Attach the hanging strip to the top with 2" countersunk screws; the strip will be temporarily removed when you hang the valance.*

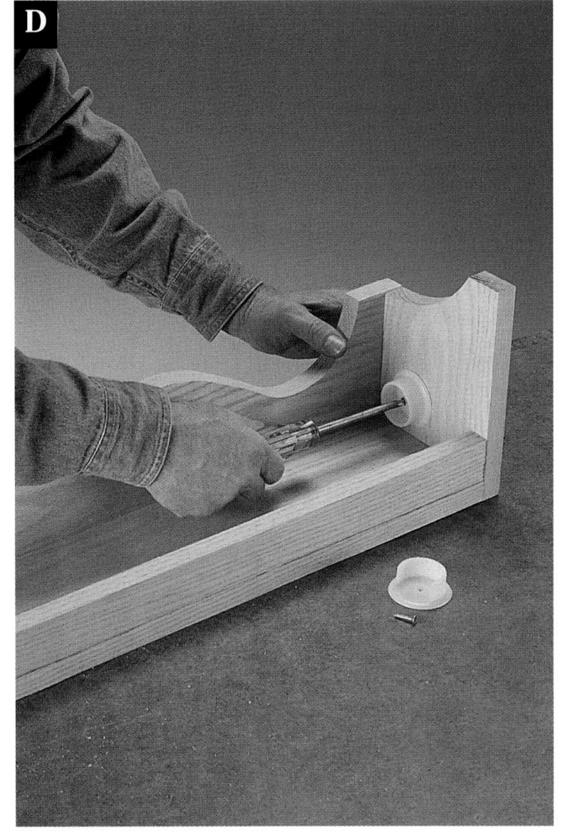

*Position the plastic rod pockets slightly forward, so they will be hidden from view when drapes are hung.*

valance, then measure and mark the desired location of the valance on the wall. Locate the studs, then position the hanging strip against the wall.

Drill mounting holes through the strip and into the wall studs, then attach the hanging strip, using 2½" wallboard screws. Position the valance on the hanging strip, flush against the wall, and mount it by reinserting the 2" screws through the top of the valance.

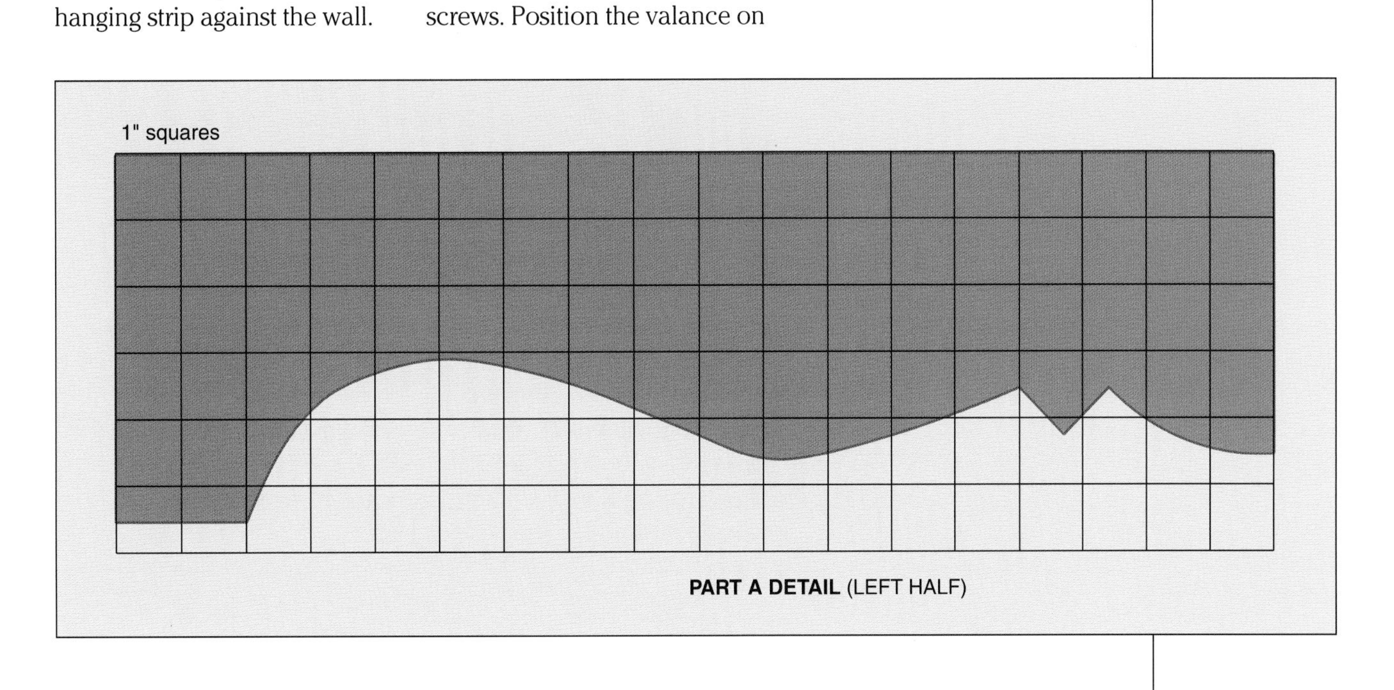


**PART A DETAIL** (LEFT HALF)

# Pipe Box

*This handsome pipe box and drawer unit makes a handy addition to any kitchen, study or workshop.*

PROJECT POWER TOOLS

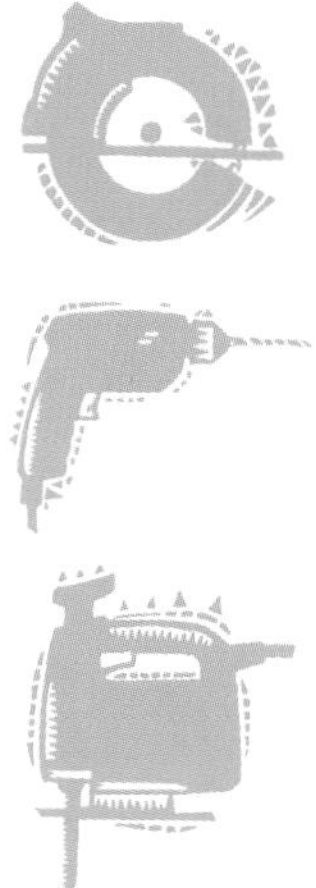

CONSTRUCTION MATERIALS

| Quantity | Lumber |
|---|---|
| 2 | ½ × 5¾" × 4' walnut |
| 1 | ⅛ × 8 × 12" hardboard |

Originally, pipe boxes were designed to hold pipes, tappers, cleaners, matches, screens and other smoking accessories. But with smoking a less popular pastime these days, you may want to use our pipe box to hold mail and stationery supplies, such as envelopes and stamps, tape and paper clips.

With its decorative lines and ornamental details, this project also offers you an opportunity to sharpen your woodworking skills. You'll find many uses for this versatile box, so you might want to make several.

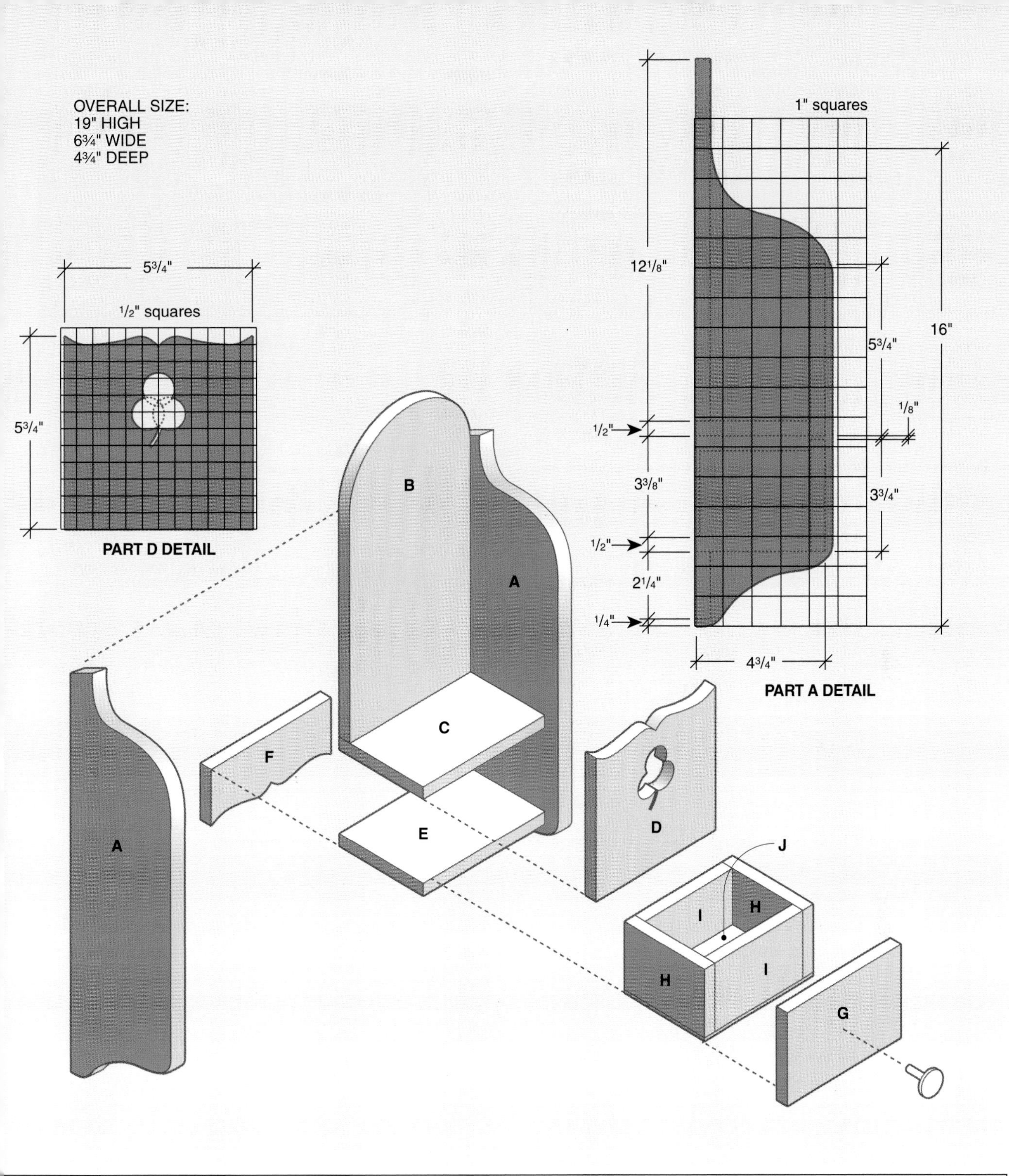


| Key | Part | Dimension | Pcs. | Material |
|---|---|---|---|---|
| **Cutting List** | | | | |
| A | Side | ½ × 4¾ × 16" | 2 | Walnut |
| B | Back | ½ × 5¾ × 12⅛" | 1 | Walnut |
| C | Upper shelf | ½ × 5¾ × 4" | 1 | Walnut |
| D | Front | ½ × 5¾ × 5¾" | 1 | Walnut |
| E | Lower shelf | ½ × 5¾ × 4" | 1 | Walnut |

| Key | Part | Dimension | Pcs. | Material |
|---|---|---|---|---|
| **Cutting List** | | | | |
| F | Apron | ½ × 5¾ × 2¼" | 1 | Walnut |
| G | Drawer front | ½ × 5½ × 3¾" | 1 | Walnut |
| H | Drawer side | ½ × 2⅝ × 3⅞" | 2 | Walnut |
| I | Drawer end | ½ × 2⅝ × 4⅜" | 2 | Walnut |
| J | Drawer bottom | ⅛ × 3⅞ × 5⅜" | 1 | Hardboard |

**Materials:** Wood glue, brads (¾", 1"), drawer pull, finishing materials.

**Note:** Measurements reflect the actual thickness of dimensional lumber.

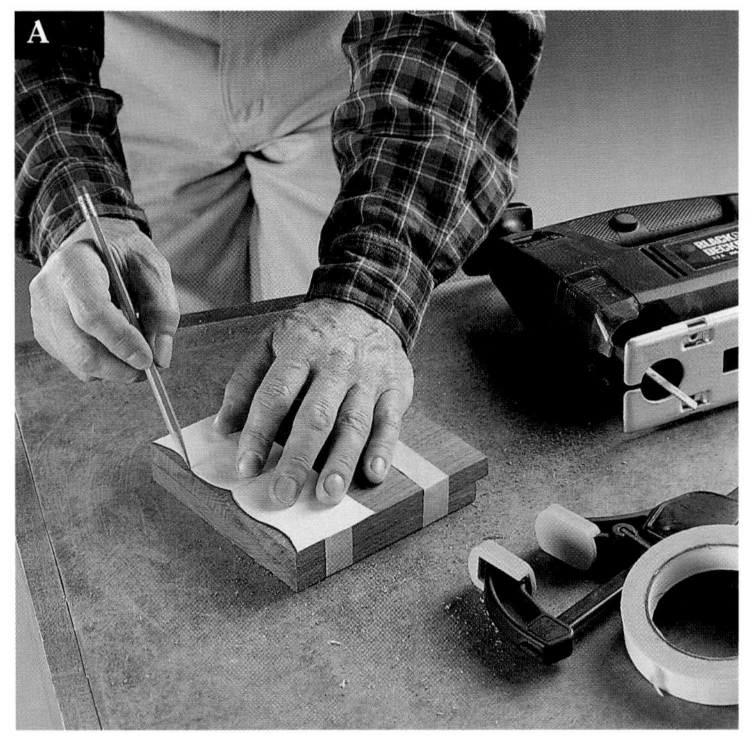


*Tape blanks for the apron and front piece together, then mark the scallops and gang-cut the profile with your jig saw.*

*Cut out the shamrock design, using a 1" spade bit for the leaves and a jig saw for the stem.*

## Directions: Pipe Box

CUT THE SIDES. First, cut out 16"-long blanks for the sides (A). Transfer the side profiles onto the material (see *Diagram*), then cut the sides using a jig saw. Sand the edges smooth.

CUT THE FRONT AND APRON. You can cut the scallop pattern on these pieces by taping the blanks together and gang-cutting with a jig saw.

Tape the blanks for the front (D) and apron (F) together with the edges flush. Transfer the scallop pattern (see *Diagram*) to the front **(photo A).** Cut both pieces with a jig saw. Remove the tape, and cut the pieces to the proper length. Sand pieces smooth. Mark the shamrock on the front (see *Diagram*), and cut out the leaf shape with a 1" spade bit **(photo B).** Sand the shamrock smooth, using a sanding dowel attached to your portable drill. Cut out the shamrock stem with your jig saw.

BUILD THE UPPER AND LOWER SHELF ASSEMBLIES. Cut the back (B) and upper shelf (C) to length. Scribe a 2⅞"-radius curve on the top of the back piece, and cut with a jig saw. Sand all cut edges smooth. Drill pilot holes and attach the upper shelf to the bottom edge of the back, using glue and 1" brads. Complete the upper shelf assembly by attaching the front piece to the front edge of the upper shelf in the same fashion **(photo C).**

Cut the lower shelf (E) to length. Complete the lower shelf assembly by drilling pilot holes, and attaching the lower shelf to the top edge of the apron, using glue and brads.

BUILD THE DRAWER. The drawer fits between the upper and lower shelves and requires careful assembly to ensure even spacing around the drawer.

First, cut the drawer parts (G, H, I, J) to size. On the back face of the drawer front, carefully measure and mark reference lines at 9⁄16" from each side, ⅜" from the top and ⅝" from the bottom. Drill pilot holes, and attach one drawer end to the drawer front, using glue and ¾" brads **(photo D).** Position the drawer sides and the remaining end, drill pilot holes, and attach the pieces with glue and 1" brads. Attach the drawer bottom with glue and brads.

### TIP

*You can increase the visual appeal of this project by making sure the grain pattern on the drawer front matches that on the front piece of the pipe box. To accomplish this, cut the two parts from the same section of wood.*

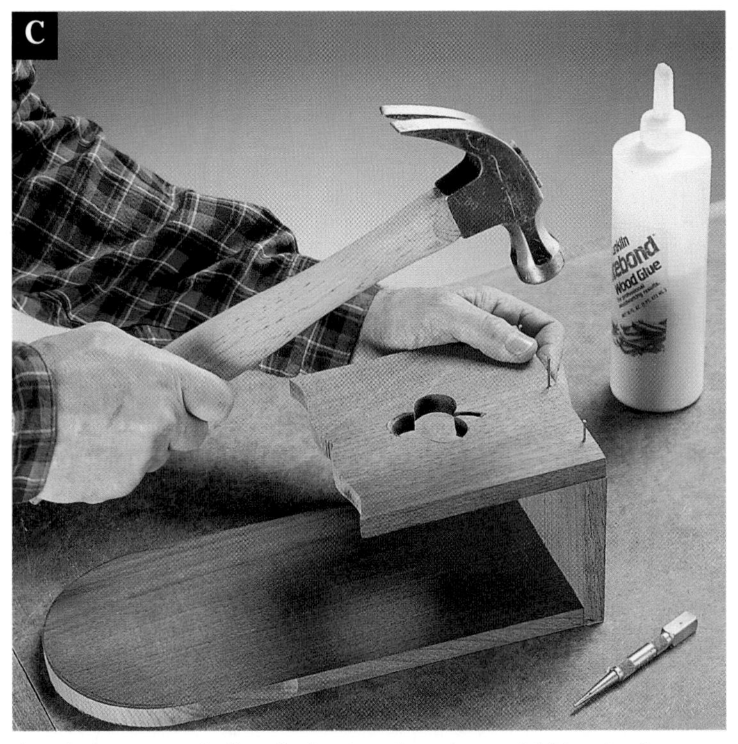


*Attach the upper shelf to the back, using glue and 1" brads, then attach the front to the shelf in the same fashion.*

*Carefully mark the back of the drawer front and attach one of the drawer ends, using glue and ¾" brads.*

*Position the sides over the shelf assemblies, using masking tape as an alignment guide.*

ASSEMBLE THE BOX. Use masking tape to mark the locations of the upper and lower shelf assemblies on the outside faces of the sides **(photo E).** Position the lower shelf assembly between the sides, then drill pilot holes and secure it with glue and 1" brads. With the drawer positioned on the lower shelf, fit the upper shelf assembly between the sides. Make sure the space above the drawer equals the space on each side (about ⅛"), then drill pilot holes and attach the upper shelf assembly with glue and 1" brads.

TIP

*You can enhance the wonderful color of walnut by applying a cherry stain before you apply a seal coat. The stain does not show up on the rich tones of the walnut, but it does give the gray/white heartwood and "bad grain" areas a very vibrant appearance. The result is a combination of rich walnut tones and beautiful "cathedral arches" of color and texture.*

APPLY FINISHING TOUCHES. Fill all visible nail holes with putty, sand the surfaces smooth and apply a finish of your choice. (We used a cherry stain and a water-based polyurethane coating.) When dry, install a decorative metal or porcelain knob on the drawer front.

You can hang your pipe box by driving a mounting screw through the back and into the wall or by attaching mounting hardware to the back of the project.

PROJECT
POWER TOOLS

# Collector's Table

*Store your fine collectibles in this eye-catching conversation piece.*

CONSTRUCTION MATERIALS

| Quantity | Lumber |
|---|---|
| 1 | ½" × 4 × 4' oak plywood |
| 1 | ½" × 4 × 8' oak plywood |
| 1 | ⅜" × 2 × 2' oak plywood |
| 2 | 1 × 2" × 6' oak |
| 4 | 1 × 3" × 8' oak |
| 2 | 1 × 4" × 6' oak |
| 1 | 2 × 2" × 6' oak |
| 3 | ½ × ½ × 30" scrap wood |
| 1 | ¼ × 18¾ × 34¾" tempered glass |

This beautiful, glass-topped collector's table is perfect for storing and displaying shells, rocks, fossils, figurines or other collectibles. It has three interchangeable drawers, so you can change the display whenever you choose—simply by rotating a different drawer into the top position under the glass.

Built from oak and oak plywood, this table gives you the opportunity to demonstrate sophisticated woodworking skills.

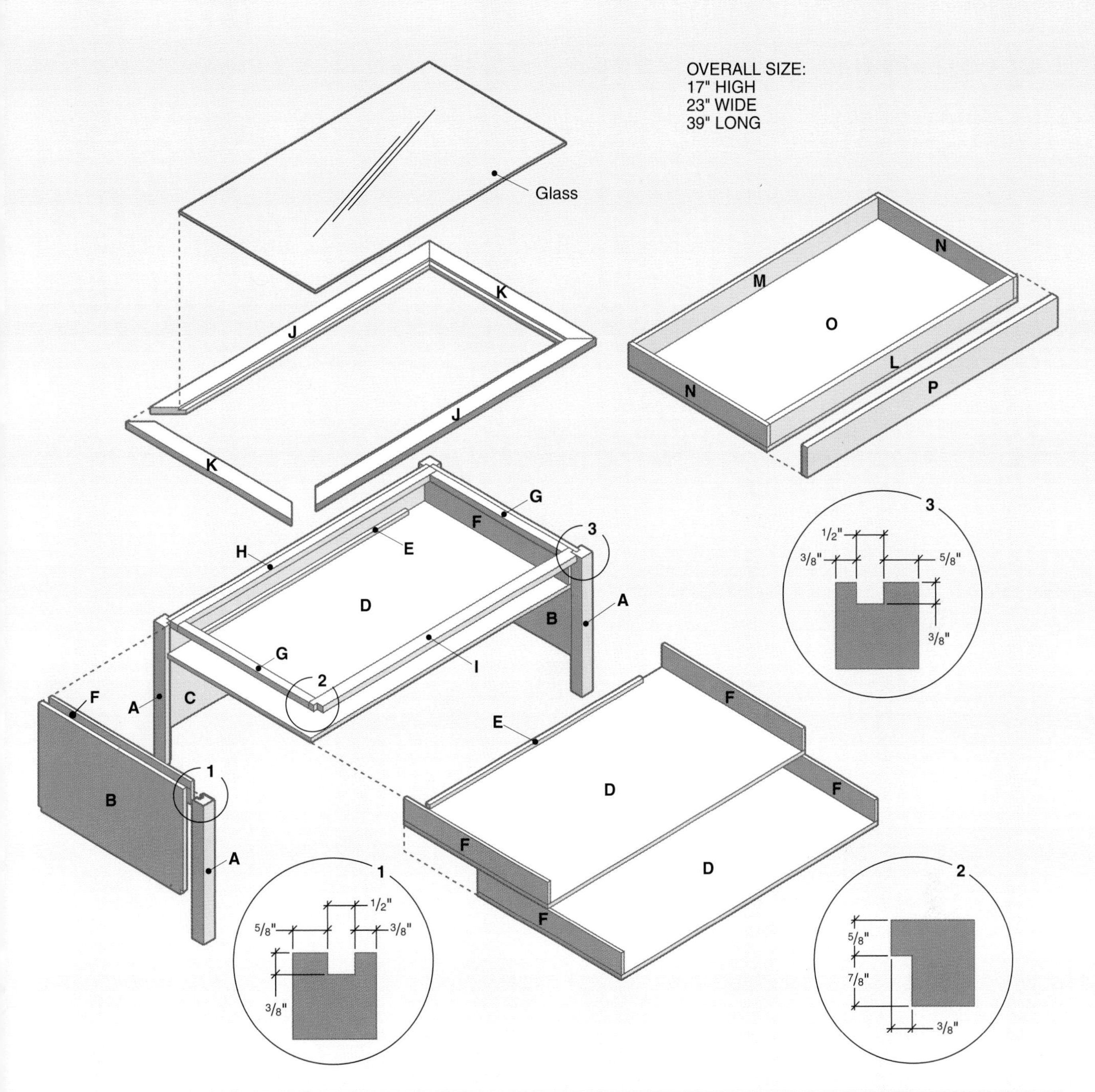


| | | Cutting List | | |
|---|---|---|---|---|
| **Key** | **Part** | **Dimension** | **Pcs.** | **Material** |
| **A** | Leg | 1½ × 1½ × 16¼" | 4 | Oak |
| **B** | End panel | ½ × 20¼ × 12½" | 2 | Oak ply. |
| **C** | Back panel | ½ × 36¼ × 12½" | 1 | Oak ply. |
| **D** | Shelf | ½ × 19⅞ × 36¼" | 3 | Oak ply. |
| **E** | Drawer stop | ½ × ½ × 30" | 3 | Scrap wood |
| **F** | Drawer guide | ⅜ × 3⅛ × 19⅞" | 6 | Oak ply. |
| **G** | End cleat | ¾ × 1 × 18¼" | 2 | Oak |
| **H** | Back cleat | ¾ × 1 × 36¼" | 1 | Oak |

| | | Cutting List | | |
|---|---|---|---|---|
| **Key** | **Part** | **Dimension** | **Pcs.** | **Material** |
| **I** | Top rail | ¾ × 1½ × 36¼" | 1 | Oak |
| **J** | Frame, long side | ¾ × 2½ × 39" | 2 | Oak |
| **K** | Frame, short side | ¾ × 2½ × 23" | 2 | Oak |
| **L** | Drawer box front | ¾ × 2½ × 34¼" | 3 | Oak |
| **M** | Drawer box back | ½ × 2½ × 34¼" | 3 | Oak ply. |
| **N** | Drawer box side | ½ × 2½ × 19½" | 6 | Oak ply. |
| **O** | Drawer bottom | ½ × 19½ × 35¼" | 3 | Plywood |
| **P** | Drawer face | ¾ × 3½ × 35¼" | 3 | Oak |

**Materials:** Wood glue, wood screws (1", 1¼"), ⅝" brads, 4d finish nails, finishing materials.

**Note:** Measurements reflect the actual thickness of dimensional lumber.

## Directions: Collector's Table

CUT THE LEGS. It is important to distinguish between the left and right legs when building this table. Each pair of front and back legs has a stopped dado to hold the end panel in place. The back legs also have a second stopped dado to support the back panel (see *Detail*). We recommend you use a router table to make these cuts.

Start by cutting the legs (A) to length. Measure and mark 12"-long dadoes on the legs. On the front legs, cut the dadoes on one face, using a ½" straight bit set to ⅜" depth. On the back legs, cut dadoes on two adjacent faces **(photo A).** Remove the waste section between the back leg dadoes with a saw and square off the dado ends, using a ½" chisel.

CUT PANELS AND ATTACH THE LEGS. Cut the end panels (B) and back panel (C) to size. Cut a ⅜ × ½" notch (see *Detail*) into the bottom corner of each panel **(photo B).** Attach a pair of legs to each end panel, using glue and brads. Set aside the back panel for attachment later.

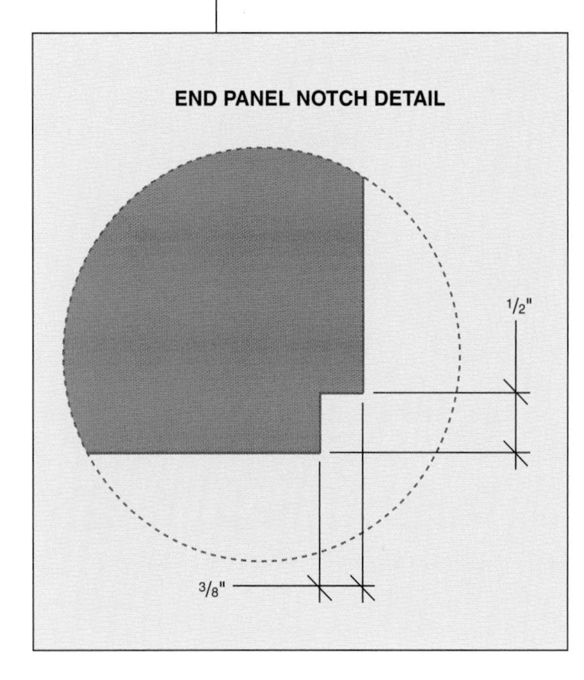

*Cut ½ × ⅜"-deep stopped dadoes in the legs, using a router table. Note the waste pieces on the back legs, which you will need to remove with a handsaw and chisel.*

*Cut a ⅜ × ½" notch out of the bottom corners of the back and side panels where they will overhang the dadoes in the legs.*

CUT THE SHELVES, DRAWER STOPS AND DRAWER GUIDES. Because the drawer stops are hidden, they can be built from any scrap ½" lumber.

Cut shelves (D), drawer stops (E) and drawer guides (F) to size. Attach the stops to the shelves with glue and brads (see *Diagram*).

*Drill pilot holes and attach the bottom drawer guides to the end panel sides, using wood glue and ⅝" brads.*

ASSEMBLE THE CABINET. Precision is crucial when assembling your collector's table. Sloppy construction will make it difficult to fit the drawers into the cabinet.

Attach the lowest drawer guides to the inside of the end panels by measuring up from the bottom edge and marking a line at 1⁄16". Position the bottom edge of the drawer guide on this line and attach with glue and brads **(photo C).**

Next, attach the back to the side assemblies by setting the back panel into the dadoes on the back legs, keeping all top edges flush. Drill pilot holes and secure the back panel with glue and brads.

Carefully turn the cabinet upside down, then drill countersunk pilot holes and attach the bottom shelf to the edge of the drawer guides with 1" wood screws. Turn the cabinet right-side-up and attach the middle shelf to the top edge of the drawer guide with 1" screws.

Rest the center drawer guides on the middle shelf and fasten to the end panels with glue and ¾" screws **(photo D).** Install the top shelf and upper guides in similar fashion.

Cut the cleats (G, H) and top rail (I) to size. Drill counterbored pilot holes and attach flush with the tops of the back and side panels, using wood glue and 1¼" screws (see *Diagram*).

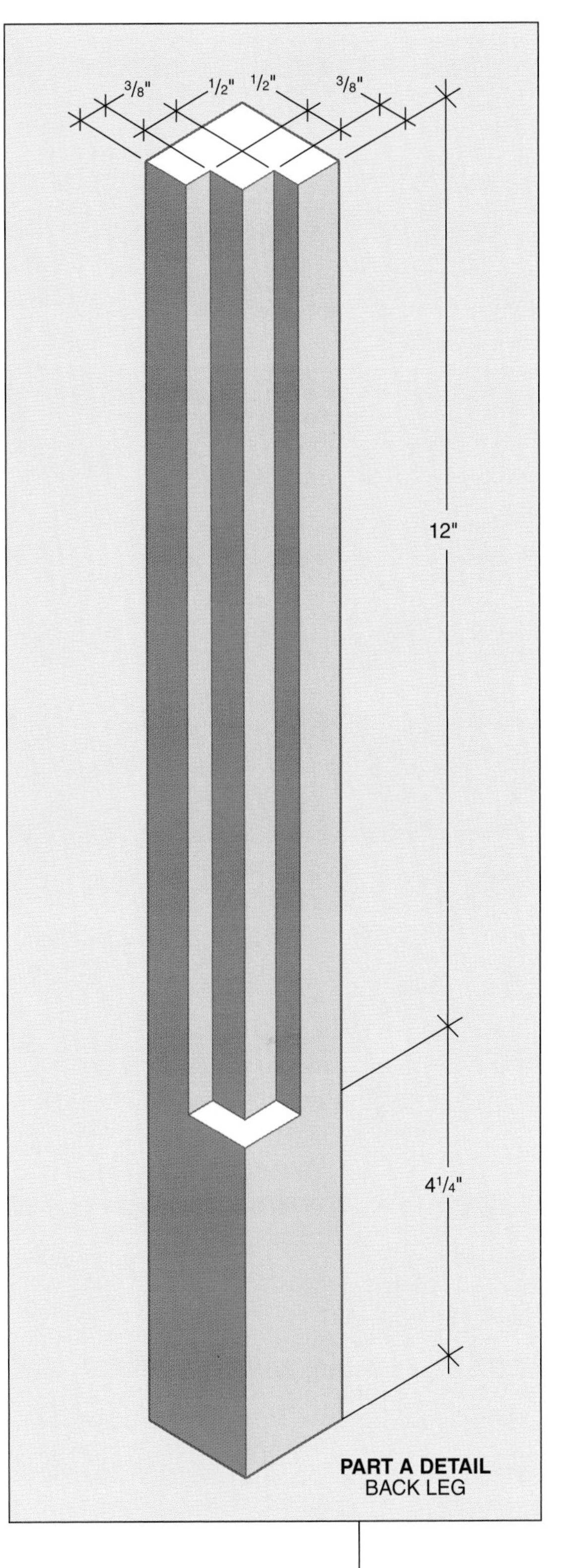


**TIP**

*Cutting stopped dadoes is a precision task. It's a good idea to practice these skills on scrap wood before you attempt this project.*

*Drill countersunk pilot holes and attach the center guides to the end panels, using wood screws.*

*Glue the miters of the top frame together, using a band clamp. Check for square by measuring diagonals.*

BUILD THE TOP FRAME. First cut the long frame sides (J) and short frame sides (K) to length, mitering the ends at 45°. Then cut a rabbet along the inside top edge of the frame pieces, using a router with a ½" straight bit set at 5⁄16" depth.

Apply glue to the mitered ends and clamp the frame together, using a band clamp **(photo E).** After the glue has dried thoroughly, center the frame over the cabinet with a ¼" overhang on all sides, drill pilot holes and attach the frame using wood glue and 4d finish nails.

MAKE THE DRAWERS. Cut the drawer box parts (L, M, N, O) to size. Assemble each drawer box by positioning the front

TIP

*For easier access to stored collections, you can attach pull knobs to the face of each drawer.*

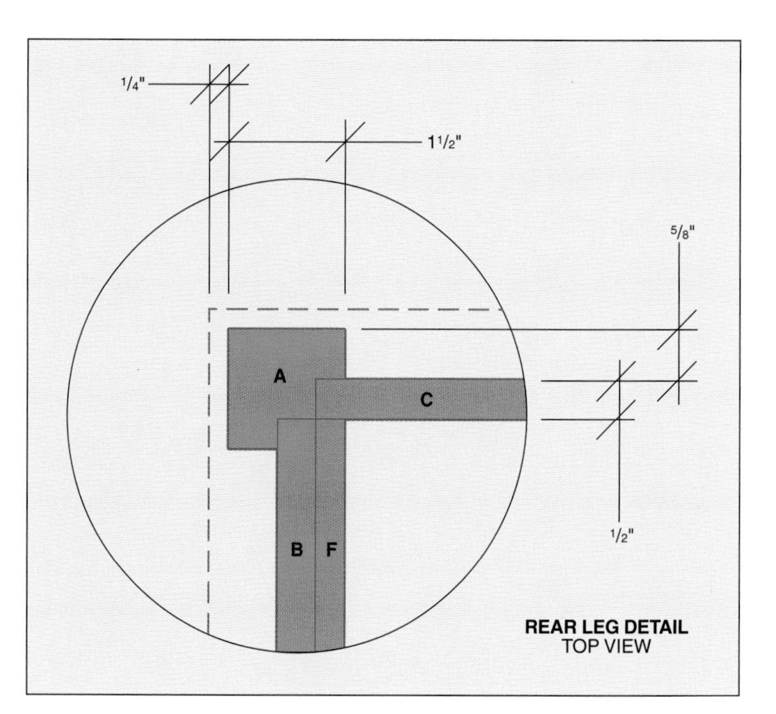

*Assemble the drawer boxes and attach the bottoms with wood glue and brads. Make sure to check for square as you build each drawer.*

*Attach drawer faces with glue, and drive wood screws through the inside of the drawer box into the drawer face.*

and back between the sides, drilling pilot holes, and attaching the pieces with glue and brads. Make sure drawer assemblies are square. Attach the drawer bottoms, using glue and brads **(photo F).**

Cut the drawer faces (P) to size. Position the drawer faces on the front of the drawer boxes, so the ends are flush and there is a 5/16" overhang at the top edge and a 3/16" overhang at the bottom. Drill pilot holes and attach by driving 1¼" wood screws from inside the drawer box fronts into the drawer face **(photo G).**

Test-fit the completed drawers in the cabinet, making sure there is ⅛" spacing between drawer faces.

FINISH THE CABINET. This project is designed to attract attention, so take care to finish it carefully.

Fill all visible nail holes with putty, then sand all surfaces smooth. Paint the insides and top edges of the drawer boxes (not the drawer faces) with flat black paint to highlight your collectibles. Stain the rest of the wood with a color of your choice (we used a light Danish walnut), and apply several coats of water-based polyurethane finish, sanding lightly between coats.

INSTALL THE GLASS TOP. It's a good idea to wait until the project is built before measuring and ordering the glass for your collector's table. Measure the length and width of the opening, and reduce each measurement by ⅛". Tempered safety glass is a good choice, especially if you have children. Seat the glass on clear, self-adhesive cushions to dampen any potential rattles.

### TIP

*If you wish, the inside of the collection drawers can be lined with black velvet to provide an ultra-elegant setting for your finest collectibles.*

PROJECT
POWER TOOLS

# Knickknack Shelf

*Add some country charm to your home with this rustic pine knickknack shelf.*

### Construction Materials

| Quantity | Lumber |
|---|---|
| 1 | 1 × 4" × 8' pine |
| 2 | 1 × 8" × 6' pine |
| 1 | 1 × 10" × 4' pine |
| 9 | ¼ × 3½" × 3' beaded pine paneling |
| 1 | ¾ × ¾" × 6' cove molding |

Country-style furniture is becoming increasingly popular throughout the world because of its honest appearance and back-to-basics preference for function over ornate styling. In fancy interior design catalogs, you may find many country shelving projects that are similar to this one in design and function. But our knickknack shelf can be built for a tiny fraction of the prices charged for its catalog cousins.

From the beaded pine paneling to the matching arcs on the apron and ledger, this knickknack shelf is well designed throughout. The shelf shown above has a natural wood finish, but it is also suitable for decorative painting techniques, like milkwash or farmhouse finishes.

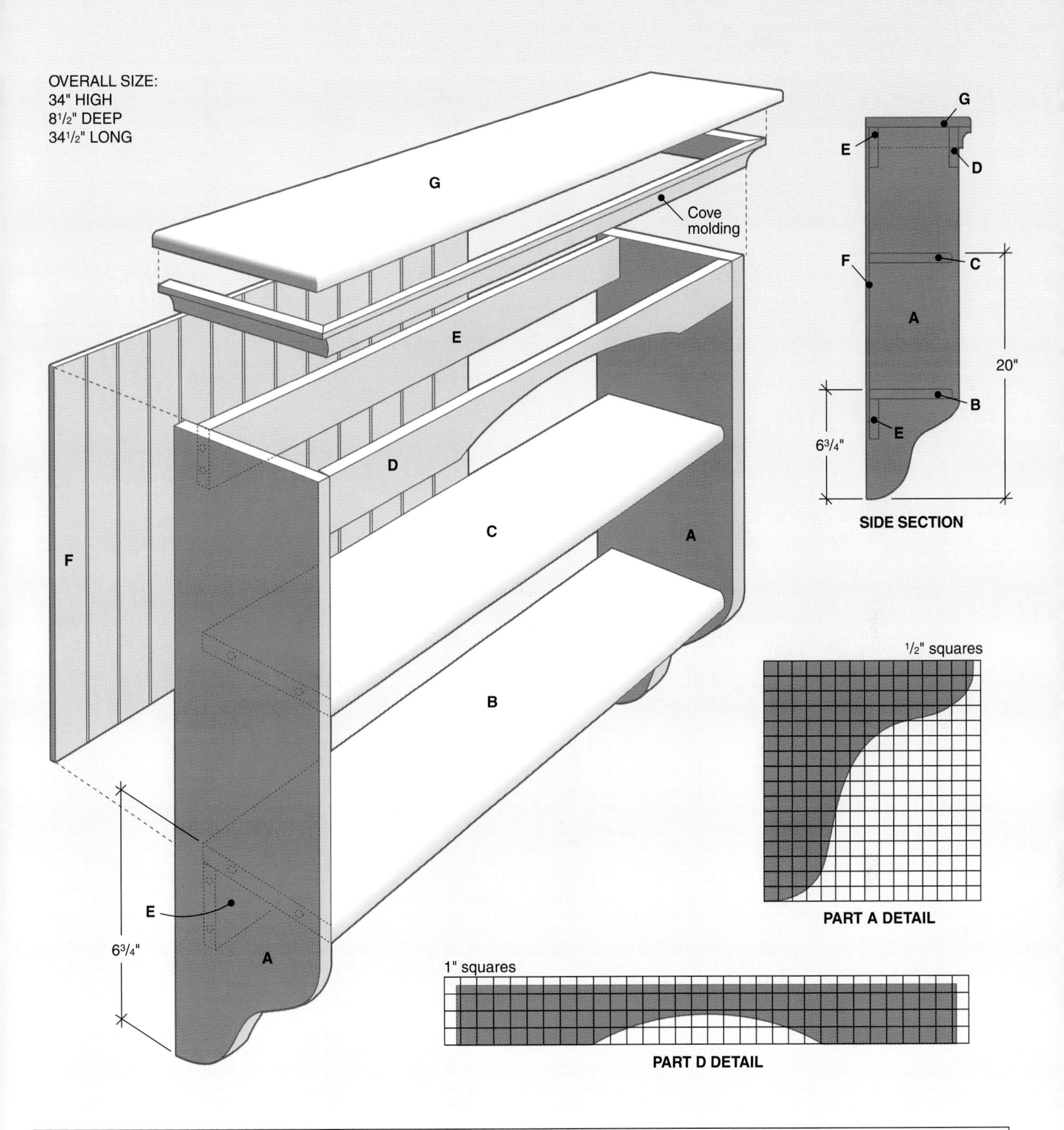


**Cutting List**

| Key | Part | Dimension | Pcs. | Material |
|---|---|---|---|---|
| A | Shelf side | ¾ × 7¼ × 33¼" | 2 | Pine |
| B | Bottom shelf | ¾ × 6¾ × 30½" | 1 | Pine |
| C | Middle shelf | ¾ × 6¾ × 30½" | 1 | Pine |
| D | Apron | ¾ × 3½ × 30½" | 1 | Pine |

**Cutting List**

| Key | Part | Dimension | Pcs. | Material |
|---|---|---|---|---|
| E | Ledger | ¾ × 3½ × 30½" | 2 | Pine |
| F | Back panel | ¼ × 3½ × 28" | 9 | Pine paneling |
| G | Cap | ¾ × 8½ × 34½" | 1 | Pine |
| | | | | |

**Materials:** Wood glue, #8 × 1½" wood screws, 3d and 6d finish nails, mushroom-style button plugs, finishing materials.

**Note:** Measurements reflect the actual thickness of dimensional lumber.

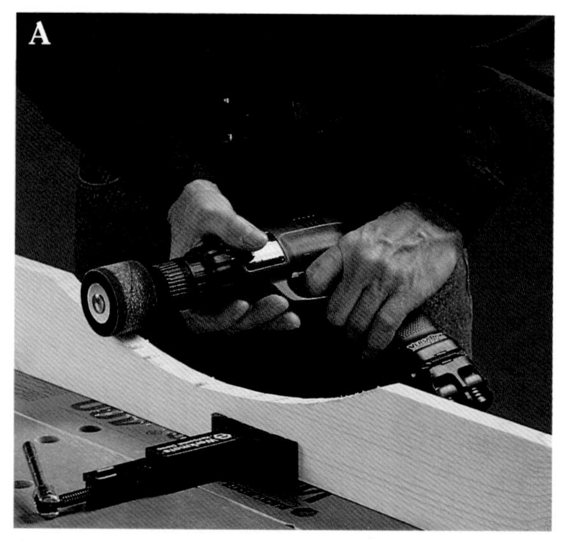

*Smooth out the jig saw cuts on the apron and ledger with a drill and drum sander.*

*Clamp the sides and ledgers in position, then fasten with glue and screws.*

## Directions: Knickknack Shelf

MAKE THE FRAME COMPONENTS. Start by cutting the sides (A) to length from 1 × 8 pine. Transfer the pattern for the sides (see *Diagram*): draw a grid with ½" squares on one of the sides, then draw the profile shown in the pattern, using the grid as a reference. Cut out the shape and smooth the cut with a drum sander attached to your drill. Trace the finished profile onto the other side, and make the cutout.

Cut the apron (D) and the ledgers (E) to length from 1 × 4 pine. Use the same technique to draw and cut out the apron and one of the ledgers (see *Diagram*). Smooth out any irregularities with a drum sander or belt sander **(photo A).**

ASSEMBLE THE FRAME. Stand the sides on their back edges and place the upper ledger between them, with the top edges flush. Place a ¼" spacer under the ledger to create a recess for the back panel. Clamp the sides and ledger together with a pipe or bar clamp.

Insert the lower ledger with its top edge 6¾" up from the bottoms of the sides, also resting on ¼" spacers. Clamp in place, and drill two counterbored pilot holes through the sides into the ends of the ledgers. Attach the sides to the ledgers with glue and wood screws **(photo B).**

INSTALL THE BACK PANEL & APRON. To make the back panel (F), we used tongue-and-groove pieces of pine wainscoting paneling, joined together and trimmed to create a 30½ × 28" panel.

Attach the back panel to the backs of the ledgers, using 3d finish nails, but no glue **(photo C).** The top of the back panel should be flush with the tops of the sides.

**TIP**

*Beaded pine paneling is easy to find in a variety of styles. One of the more economical ways to purchase it is in prepackaged cartons, usually containing 10 square feet of ¼"-thick tongue-and-groove pine with face grooves that are 4" apart.*

*Fasten the tongue-and-groove beaded pine panel pieces to the ledgers with 3d finish nails.*

Fasten the apron across the top front of the sides with wood glue and 6d finish nails. Be sure to keep the top edge of the apron flush with the tops of the sides.

BUILD & INSTALL THE SHELVES. Cut the bottom shelf (B) and middle shelf (C) to size from 1 × 8 pine. Using a router with a ⅜" piloted roundover bit, round over the top and bottom edges on the fronts of the shelves **(photo D).**

Clamp the bottom shelf in place on top of the lower ledger, keeping the back edges flush. Drill counterbored pilot holes, and attach the shelf with glue and wood screws. Install the middle shelf using the same procedure **(photo E).**

ATTACH THE CAP & COVE. Cut the cap (G) to size from 1 × 10 pine. Using a router with a ⅜" roundover bit, shape the top and bottom edges of the ends and front. Place a bead of glue along the top edges of the shelf sides, apron and ledgers. Position the cap on top of the shelf assembly, overlapping 1¼" on each end and at the front. Nail the cap in place with 6d finish nails.

Cut the pine cove molding to the appropriate lengths with mitered corners, and attach just below the bottom of the cap (see *Diagram*). Fasten in place with glue and 3d finish nails **(photo F).**

APPLY FINISHING TOUCHES. Decide where to hang the knickknack shelf, and drill counterbored pilot holes in the upper ledger for mounting screws. Scrape off any excess glue, then finish-sand. Install mushroom-style button plugs in all counterbores, then apply the finish. We chose to finish our knickknack shelf with light oak stain and a satin-gloss polyurethane topcoat.

*Use a router with a ⅜" piloted roundover bit to shape the front edges of the shelves.*

*Drill counterbored pilot holes for the shelves, then attach with glue and 1½" wood screws.*

*Attach the cove molding with glue and 3d finish nails. Hold the nails with needlenose pliers when nailing in hard-to-reach areas.*

PROJECT
POWER TOOLS

# Plant Stand

*Wherever you need it, this cedar plant stand provides a setting in which your plants will thrive.*

CONSTRUCTION MATERIALS

| Quantity | Lumber |
|---|---|
| 1 | 1 × 8" × 8' cedar |
| 1 | 1 × 8" × 6' cedar |
| 1 | 1 × 4" × 4' cedar |
| 1 | 1 × 2" × 4' cedar |
| 1 | 1 × 2" × 6' cedar |

Though it looks simple, our plant stand is designed to serve many needs. On its tall top shelf, small plants will catch plenty of sunlight. The wide lower shelf, reinforced with cleats and supports, will safely hold large potted plants. Made of cedar, the plant stand looks great indoors, but it can also be placed on a patio or deck, where it will age naturally to a weathered gray. To enhance its rustic beauty, we created decorative diamond cutouts by gang-cutting notches in the side pieces, then rounded the shelf corners with a belt sander.

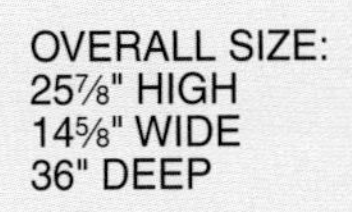


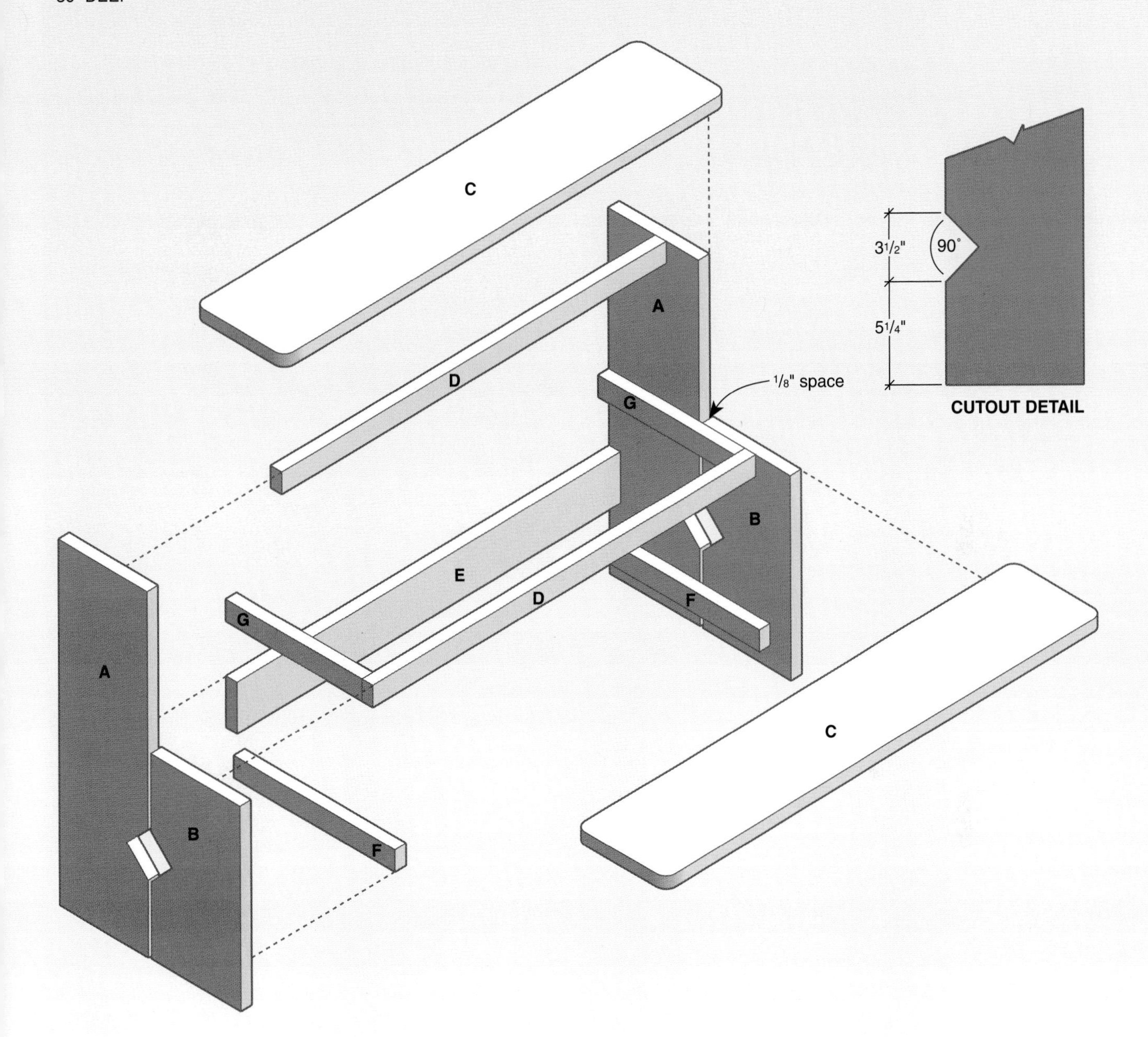


**Cutting List**

| Key | Part | Dimension | Pcs. | Material |
|---|---|---|---|---|
| A | Long side | ⅞ × 7¼ × 25" | 2 | Cedar |
| B | Short side | ⅞ × 7¼ × 14" | 2 | Cedar |
| C | Shelf | ⅞ × 7¼ × 36" | 2 | Cedar |
| D | Support | ⅞ × 1½ × 31¼" | 2 | Cedar |

**Cutting List**

| Key | Part | Dimension | Pcs. | Material |
|---|---|---|---|---|
| E | Stretcher | ⅞ × 3½ × 31¼" | 1 | Cedar |
| F | Long cleat | ⅞ × 1½ × 13" | 2 | Cedar |
| G | Short cleat | ⅞ × 1½ × 11" | 2 | Cedar |
| | | | | |

**Materials:** Waterproof wood glue, yellow deck screws (2", 1½"), finishing materials.

**Note:** Measurements reflect the actual thickness of dimensional lumber.

*Mark ¾"-radius curves on the corners of the shelves, and grind them down to the lines using a belt sander.*

*Use a jig saw to gang-cut triangular notches in the side pieces, ensuring symmetrical cutouts.*

## Directions: Plant Stand

MAKE THE SHELVES. Both of the shelves have rounded corners to soften the look of the project.

Start by cutting the shelves (C) to length from 1 × 8 cedar. To make the rounded corners, use a compass to draw ¾"-radius curves on the corners of both shelves. Clamp a belt sander perpendicular to your worksurface and round the corners by sanding to the lines **(photo A).**

MAKE THE SIDES. Each side has a triangle cutout that forms a diamond when the long and short sides are joined. To ensure that the diamond is symmetrical, position the long and short sides together and gang-cut the notches.

Cut the long sides (A) and short sides (B) to length. Stack a long and short side together with the edges and bottoms flush, and secure them with box tape. Mark a triangular notch on one edge (see *Diagram*); the corner of a framing square works well for marking this 90° notch.

Clamp the pieces to your worksurface and cut the notch with a jig saw. Cut matching notches on the other two side pieces, using the same procedure **(photo B).** Sand the cut edges smooth.

ATTACH THE CLEATS. Because the plant stand will inevitably come into contact with water, we used waterproof glue throughout the assembly.

Cut the long cleats (F) and the short cleats (G) to length. Lay out a long side and a short side so the edges are flush and the diamond cutout is aligned. Place ⅛" spacers between the sides. Position a short cleat

### TIP

*Most cedar is rough on one side. When assembling, be sure that exposed surfaces are consistent in texture. For this project, we chose to have the smooth sides facing outward.*

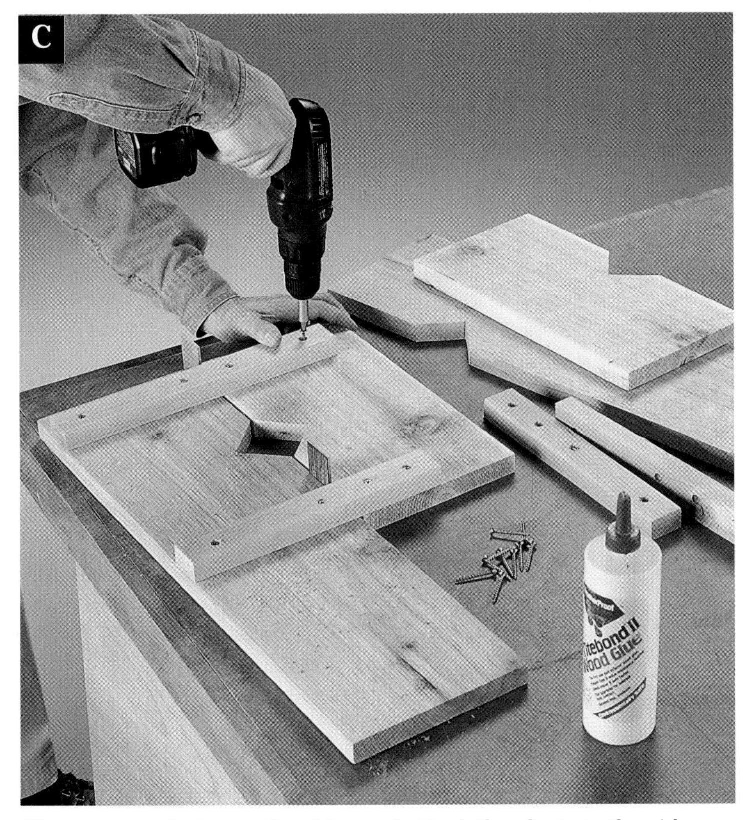


*Place spacers between the sides and attach the cleats to the sides with wood glue and deck screws.*

*Position the supports and stretcher between the sides and join with glue and deck screws.*

flush with the top edge of the short side and the back edge of the long side. Drill countersunk pilot holes through the short cleat into the sides. Attach the cleat with waterproof glue and 1½" yellow deck screws. Position a long cleat 1½" up from the bottom edges of the sides with the back edges flush. Drill countersunk pilot holes through the cleats into the sides and attach with waterproof glue and 1½" screws **(photo C).** Remove the spacers and repeat the process for the other sides.

ATTACH THE SUPPORTS AND STRETCHER. Like the cleats, the supports and stretcher provide stability and ensure that the shelves will not bend under the weight of heavy plants.

Cut the supports (D) and stretcher (E) to length. Position the stretcher between the sides so the bottom edge is 5¼" from the bottom and the back edges are flush. Drill countersunk pilot holes through the sides into the stretcher. Attach with waterproof glue and 2" deck screws. Position the lower support against the ends of the short cleats. Drill countersunk pilot holes and attach with waterproof glue and 2" deck screws. Position the top support between the long sides and connect with waterproof glue and 2" deck screws driven through countersunk pilot holes **(photo D).**

ATTACH THE SHELVES. Center the top shelf from side to side and drill countersunk pilot holes to connect the shelf to the side pieces. Apply glue and drive 2" deck screws through the holes. Repeat this process to attach the bottom shelf.

APPLY FINISHING TOUCHES. Scrape off any excess glue. Sand with medium-grit sandpaper to break the edges and smooth any rough spots.

**TIP**

*Cedar will naturally gray as it dries and weathers. If you want your cedar project to retain its original color, apply a clear wood sealer.*

PROJECT
POWER TOOLS

# Shadow Box

*Display your prized collection in this attractive shadow box.*

### Construction Materials

| Quantity | Lumber |
|---|---|
| 1 | 1 × 2" × 8' oak |
| 1 | 1 × 3" × 6' oak |
| 1 | 1 × 3" × 4' oak |
| 3 | ½ × 2½" × 5' oak |
| 1 | ¼" × 2 × 4' oak plywood |
| 1 | ⅛ × 17½ × 27½" Plexiglas® |

Collectibles shouldn't be kept on a table or scattered around a room where they might be ignored. Our shadow box allows you to organize and display your collection so admirers can fully appreciate it. The shadow box has five levels and two different compartment sizes to accommodate items of varying shapes. Its door and Plexiglas panel keep valuable contents dust-free. Hidden barrel hinges and dowel joints on the door eliminate exterior hardware that can distract attention from your treasures. Mount this shadow box to your wall and give your collection the display it deserves.

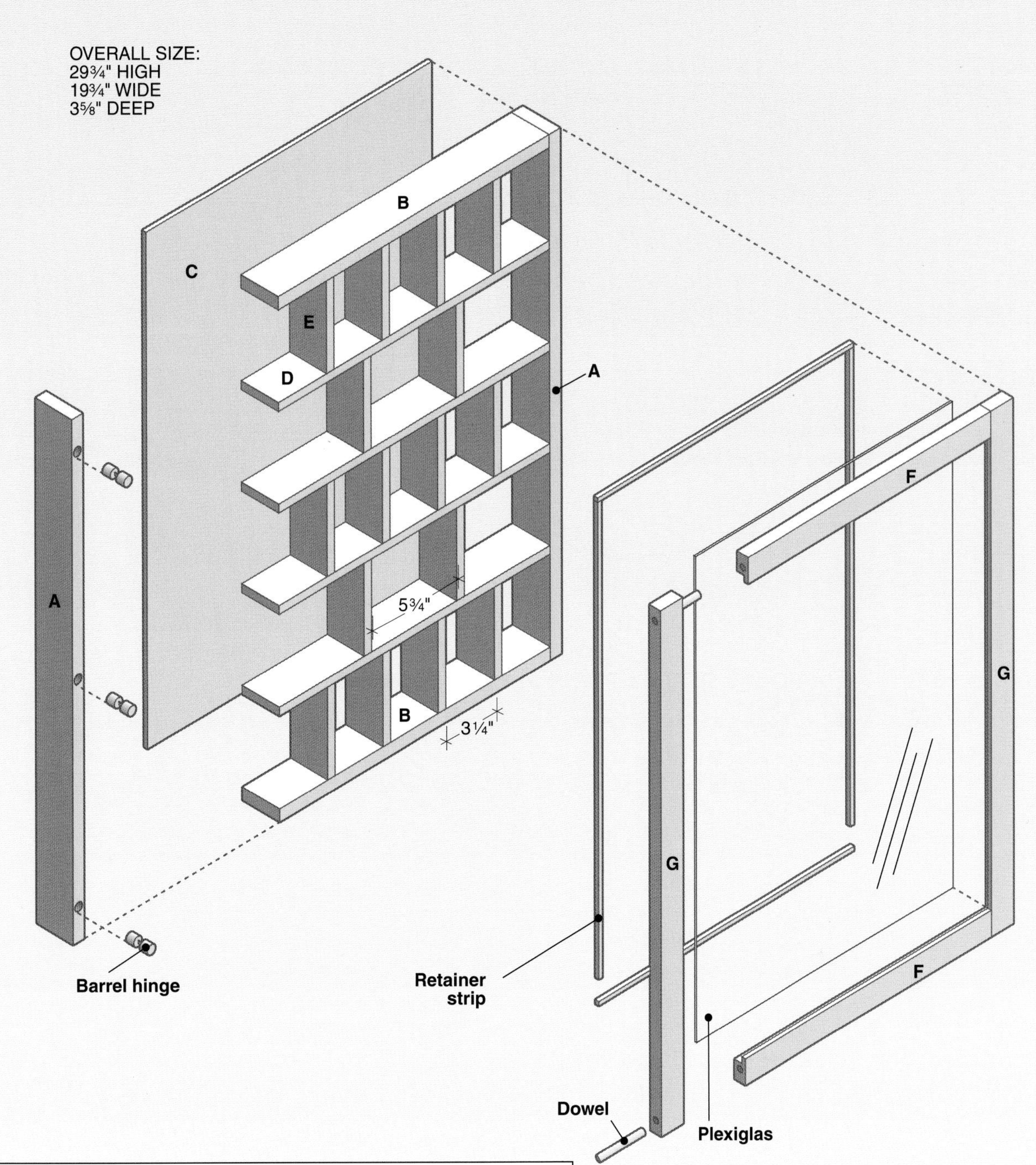


### Cutting List

| Key | Part | Dimension | Pcs. | Material |
|---|---|---|---|---|
| A | Side | ¾ × 2½ × 29¾" | 2 | Oak |
| B | Top/bottom | ¾ × 2½ × 18¼" | 2 | Oak |
| C | Back | ¼ × 19¼ × 29¼" | 1 | Oak plywood |
| D | Shelf | ½ × 2½ × 18¼" | 4 | Oak |
| E | Divider | ½ × 2½ × 5¼" | 16 | Oak |
| F | Door rail | ¾ × 1½ × 16¾" | 2 | Oak |
| G | Door stile | ¾ × 1½ × 29¾" | 2 | Oak |

**Materials:** Wood glue, 2" wallboard screws, #6 wood screws (1", 1½"), ⅝" brads, ⅜" oak plugs, ⅜ × 3" dowels, 3 hidden barrel hinges for ¾" wood, panel retainer strip (8'), finishing materials.

**Note:** Measurements reflect the actual thickness of dimensional lumber.

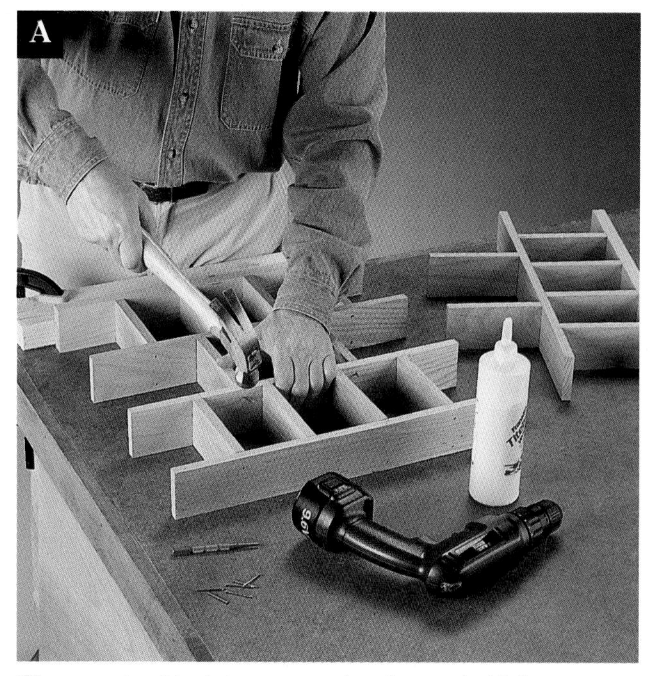

*Clamp a stop block to your worksurface to hold the assembly steady as you attach the dividers to the shelves.*

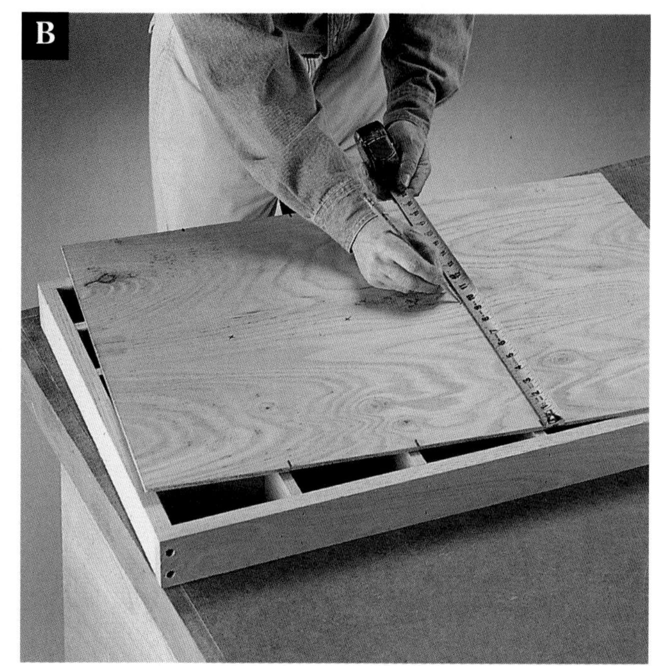

*Draw reference lines on the back panel to show the locations of the shelf centerlines.*

## Directions: Shadow Box

ASSEMBLE THE SHELF GRID. The shelf grid consists of five sections—three with smaller compartments, two with larger spaces. Build the smaller sections first, then join them together to create the larger compartments.

Cut the top and bottom (B) and shelves (D) to size, and sand the pieces. On the edges of these horizontal pieces, mark the positions of the vertical dividers that will form the smaller compartments (see *Diagram*).

Cut the dividers (E) to size using a power miter box (clamp a stop block to the saw to ensure that all dividers are exactly the same length). Sand the cut edges smooth. Join the dividers to the horizontal pieces to form the three small-compartment sections, using glue and ⅝" brads end-nailed through pilot holes.

On the edges of the shelves, mark the locations of the vertical dividers that will form the large compartments. Complete the shelf grid by drilling angled pilot holes and attaching the dividers with glue and brads **(photo A).** Set all nail heads.

ATTACH THE SIDES AND BACK. Cut the sides (A) to size and sand smooth. Position the sides against the shelves so the ends are flush with the top and bottom, and hold the pieces together with bar clamps. Drill counterbored pilot holes through the sides into the top and bottom, then attach the sides with glue and 1½" screws.

Because the back will support the weight of the shadow box once the unit is hung, it must be securely attached.

Cut the back (C) to size and sand smooth. Lay the back on the shelf assembly and mark it to show the centerlines of the horizontal shelves **(photo B).** Drill countersunk pilot holes through the back into the shelves, sides, top and bottom, and attach the back with 1" screws.

MAKE THE DOOR. Cut the rails (F) and stiles (G) to size and sand smooth. On your worksurface, position the rails between the stiles so the edges are flush. At each corner, drill a ⅜ × 3"-deep hole through the stile and into the rail. Join the

### TIP

*Barrel hinges come in different metric sizes for use with wood of varying thicknesses. For ¾" wood, use a 14mm drill bit for the holes. You'll also need a special screwdriver to lock the barrels into place; it can be purchased from the same source as the hinges.*

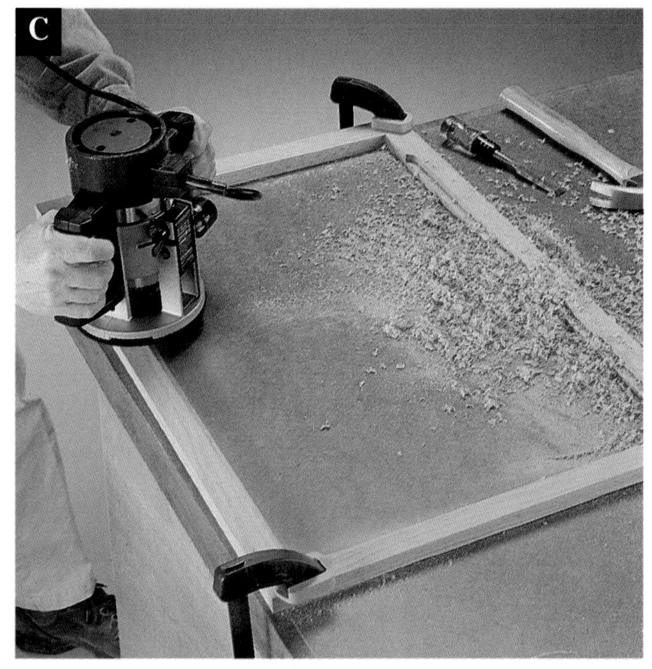

*Route a ⅜ × ½"-deep rabbet around the inside of the door frame to hold the Plexiglas panel.*

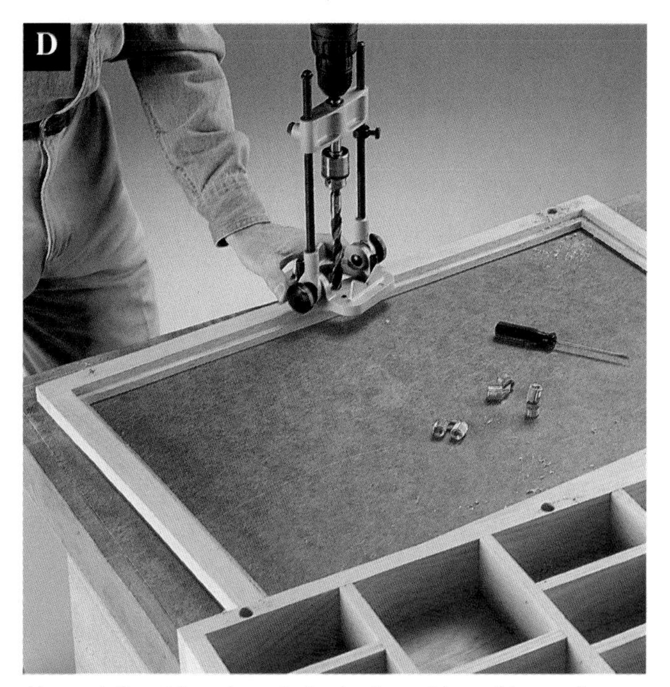

*Use a drill guide to bore holes in the cabinet sides and door stiles for the barrel hinges.*

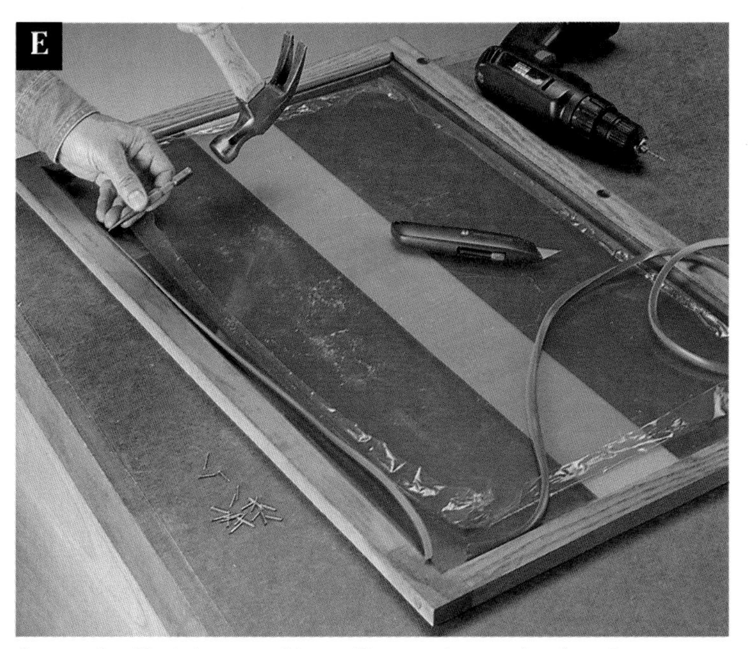

*Secure the Plexiglas panel by nailing retainer to the door frame.*

rails and stiles with glue and dowels. Check for square, adjust as necessary and secure the pieces with bar clamps until the glue dries.

The inner edge of the door frame is rabbeted to hold the Plexiglas panel. To ensure smooth rabbets, cut them with two passes of the router.

Clamp the door assembly facedown on your worksurface. Rout a ⅜"-wide × ½"-deep rabbet along the inside of the door frame **(photo C).** Use a sharp chisel to square off the corners.

ATTACH THE DOOR. Measure and mark three aligned holes for the barrel hinges on the back of the door frame and the front edge of the left side. Use a drill guide, and follow the hinge manufacturer's depth specifications to drill the holes **(photo D).**

APPLY FINISHING TOUCHES. Insert ⅜" glued oak plugs into the counterbored screw holes and sand flush. Fill nail holes with wood putty. Finish-sand the project and apply a finish of your choice (we used a hickory stain). Place the Plexiglas in the rabbet and attach panel retainer strips with ⅝" brads driven through angled pilot holes **(photo E).** Set the nail heads. Install the barrel hinges.

To hang the shadow box on a wall, drive 2" wallboard screws through the back, into wall studs.

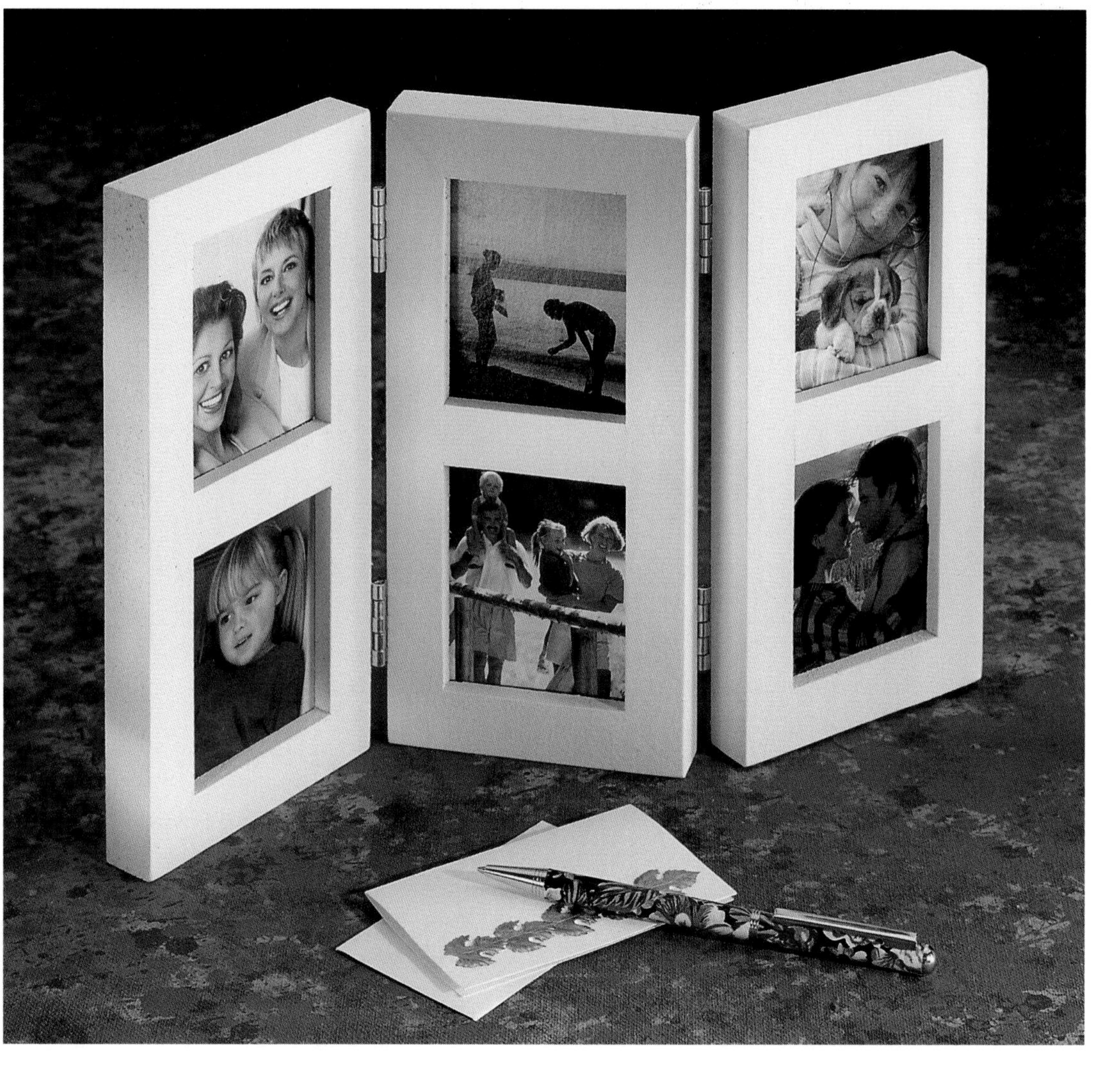

PROJECT
POWER TOOLS

# Picture Frame

*Here's a great gift project that displays your fondest memories and your finest craftsmanship.*

**Construction Materials**

| Quantity | Lumber |
|---|---|
| 1 | ¾ × ¾" × 8' birch |
| 1 | ¼ × 2 × 12" birch |
| 1 | ⅛ × 8 × 8" hardboard |
| 3 | ⅛ × 2½ × 6½" glass |

This project makes an ideal gift for grandparents, uncles, aunts, friends and neighbors. In fact, once word starts to spread that you're making these attractive tri-fold picture frames, you'll become a busier woodworker, trying to keep up with all the requests. These paintable frames are designed to hold six 2¾ × 2¼" photographs, but you can make larger frames by proportionately increasing the dimensions and applying the same construction methods. The end frames have hinges mounted on opposite sides so you can open the picture frames to create a freestanding, S-curved display of your most treasured memories.

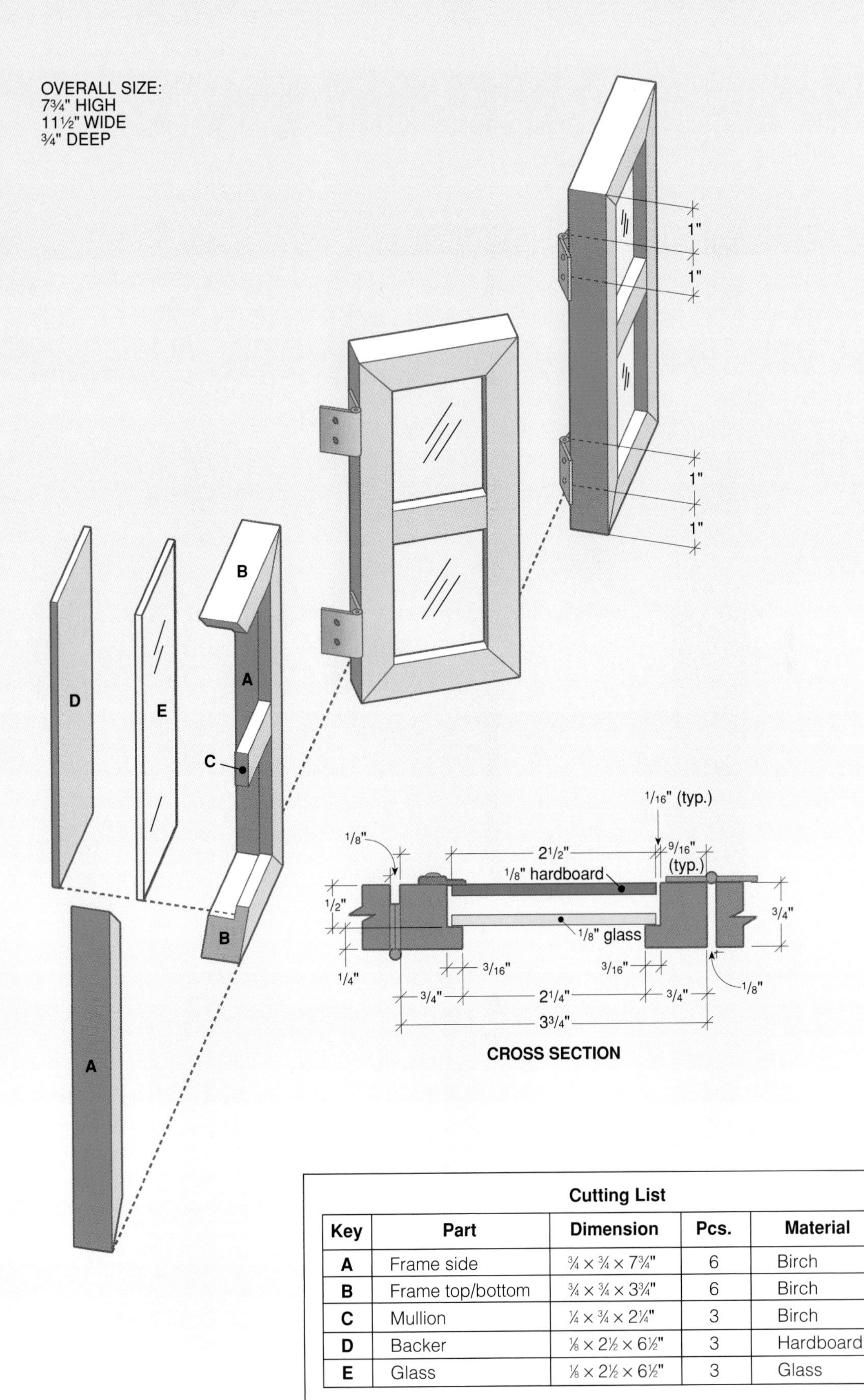


| Cutting List | | | | |
|---|---|---|---|---|
| **Key** | **Part** | **Dimension** | **Pcs.** | **Material** |
| **A** | Frame side | ¾ × ¾ × 7¾" | 6 | Birch |
| **B** | Frame top/bottom | ¾ × ¾ × 3¾" | 6 | Birch |
| **C** | Mullion | ¼ × ¾ × 2¼" | 3 | Birch |
| **D** | Backer | ⅛ × 2½ × 6½" | 3 | Hardboard |
| **E** | Glass | ⅛ × 2½ × 6½" | 3 | Glass |

**Materials:** Wood glue, 4 brass hinges (1 × ½"), 12 retaining clips, finishing materials.

**Note:** Measurements reflect the actual thickness of dimensional lumber.

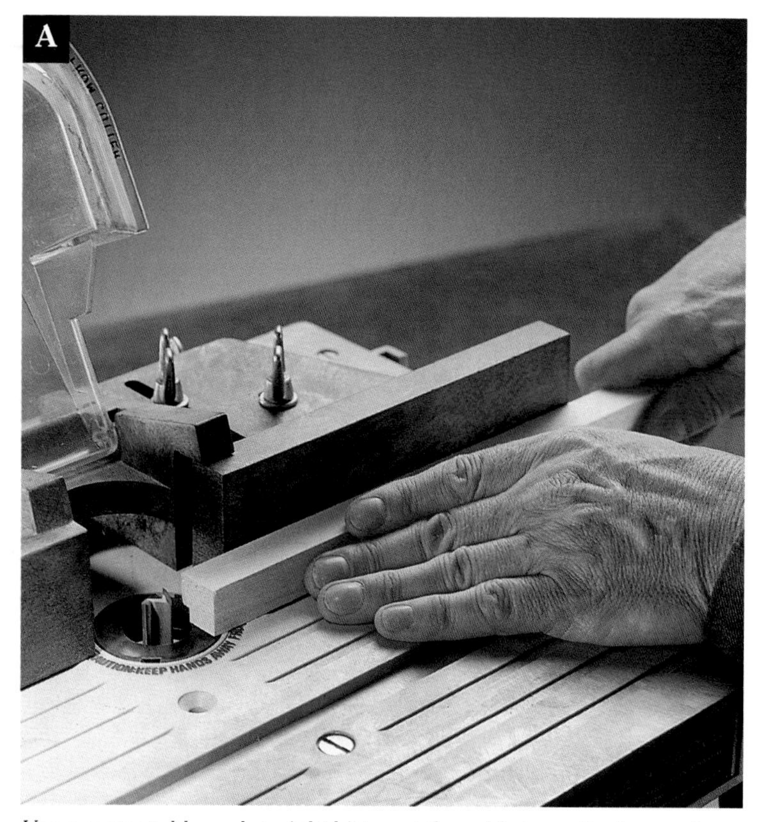


*Use a router table and straight bit to cut the rabbets on the frame pieces.*

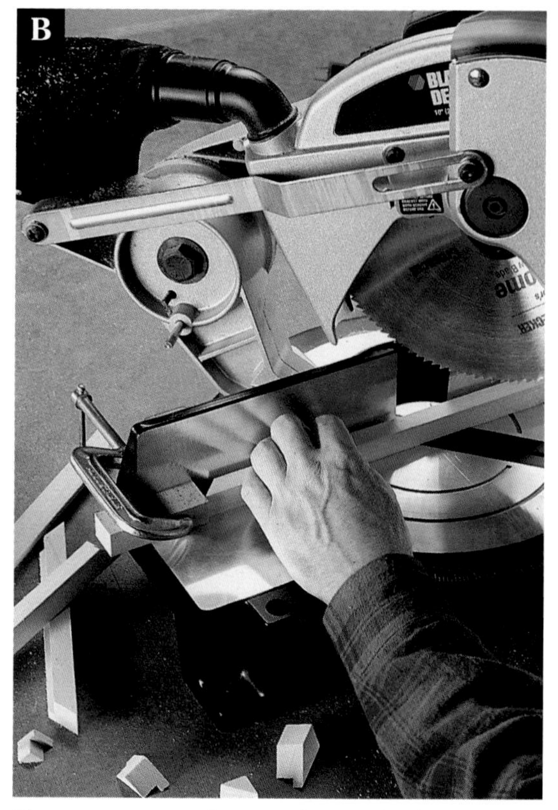

*Use a power miter box to miter-cut the frame pieces to length.*

**TIP**

*We built our frames from birch, which can be stained, painted or left unfinished, but any hardwood is suitable. Exotic woods will give your picture frame unique appeal.*

## Directions: Picture Frame

CUT THE PARTS. To cut the rabbets on the inside faces of the frame pieces, we recommend using a router table and a straight router bit. Make test cuts on scrap material to ensure accurate rabbets.

Set the fence of the router table to 3⁄16". Cut the rabbets on the frame pieces by making multiple passes, gradually extending the depth of the cut until you achieve the ½" depth **(photo A).** This technique creates a cleaner rabbet and reduces the risk of tearouts.

Using a power miter box, carefully cut the frame sides (A) and tops/bottoms (B) to length, mitering the ends at 45° **(photo B).** Clamp a stop block to the fence of the miter saw to ensure the pieces are cut to the exact same length.

ASSEMBLE THE FRAMES. Glue the frame sides, bottoms and tops together, and secure with band clamps **(photo C).** Make sure the frames are square. Leave excess glue until it hardens, then gently remove the dried glue with a sharp chisel.

Rip-cut ¼" stock to ⅞" width, then cut the mullions (C) to length. Mark a "sand-to" line on the mullions. Clamp a belt sander onto your worksurface in a horizontal position, then grind down the mullions to the marked lines **(photo D).** Test-fit the mullions in the frames. Attach with glue, and clamp until dry.

ALIGN THE HINGES. Measure, mark and drill pilot holes for attaching the hinges. Make sure the hinge barrels face front in

*Join the frame pieces with glue and secure them with band clamps until the glue dries.*

D

*Use a belt sander to grind down the mullions to the "sand-to" lines.*

one reveal and face back in the other reveal. The hinges with front-facing barrels are attached to the sides of the frames; the other hinges are mounted on the back of the frames. Install the hinges, then remove them from the frames (they will be reinstalled after the finish is applied).

APPLY FINISHING TOUCHES. Finish-sand the project, then finish as desired. If you want to highlight the wood grain, you can use an aniline dye to stain the wood, provided the joints are tight and clean. Otherwise, paint is always a good option.

After the finish dries, reinstall the hinges **(photo E).** Cut the hardboard backers (D) to size. Insert the glass, photographs and backers. Add a layer of cardboard as a spacer, then secure with retaining clips.

*One pair of hinges is attached with the barrels facing the front; the other hinges are attached so the barrels face the back.*

PROJECT
POWER TOOLS

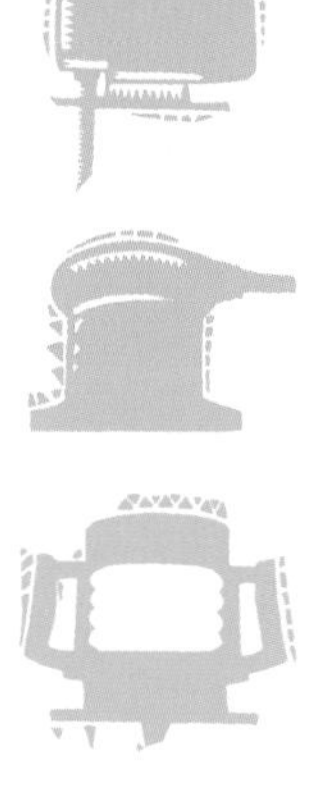

# Plate & Spoon Rack

*This easy-to-build rack displays your collectibles and shows off your woodworking ability.*

CONSTRUCTION MATERIALS

| Quantity | Lumber |
|---|---|
| 1 | 21⁄32 × 24 × 47½" ponderosa pine panel |

This decorative fixture, made from edge-glued ponderosa pine panels, has features you'll appreciate as your collection continues to grow. The back features three heart cutouts, and scallops along the top edge accentuate your most prized plates. The plate shelf has a groove cut into it to help stabilize up to three full-size plates, while the sides are curved to soften the edges and to better display the plates. The spoon rack sits in front of a curved background that adds interest to the unit, while the rack itself has notches that hold up to seventeen collectible teaspoons in full view.

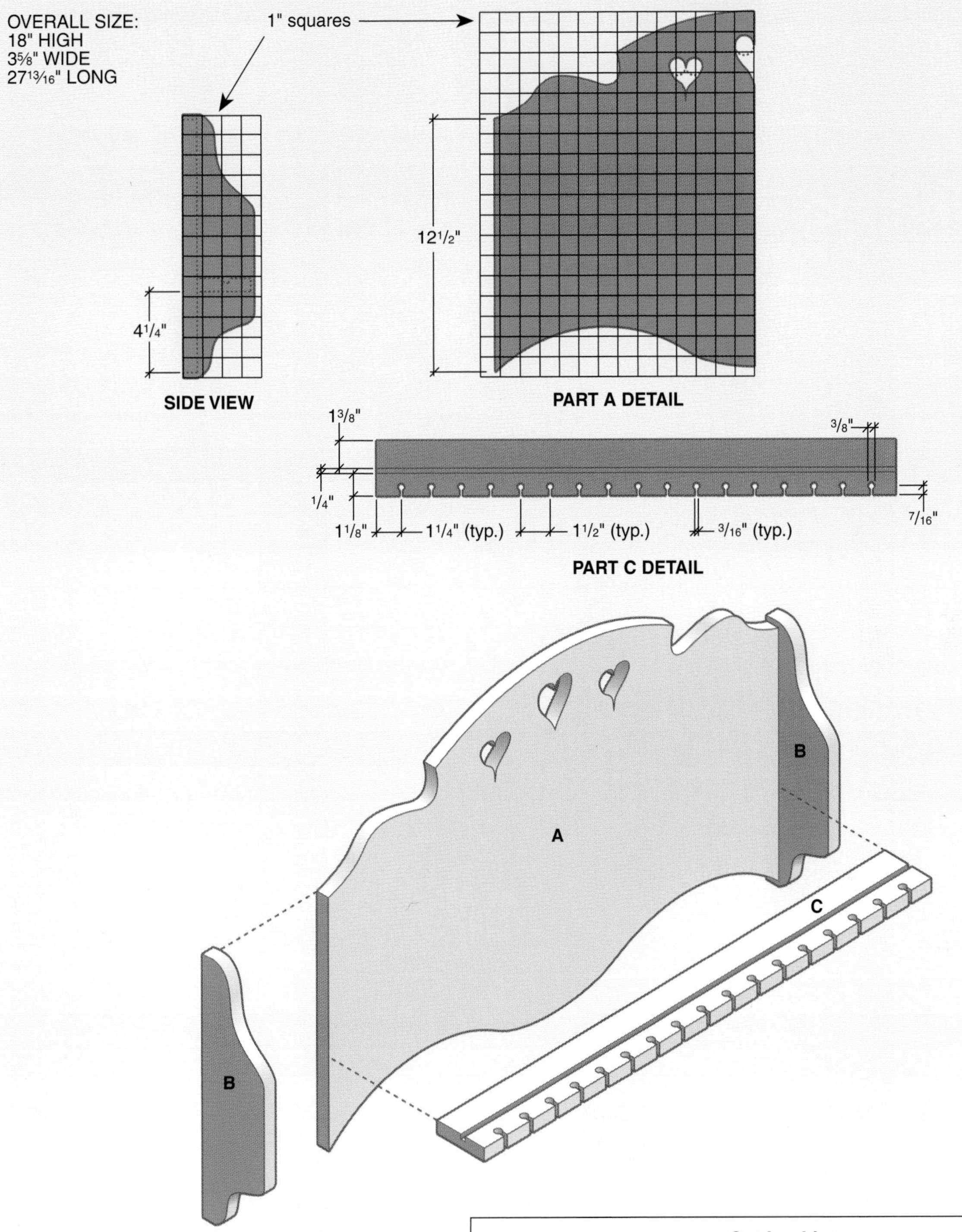


**Cutting List**

| Key | Part | Dimension | Pcs. | Material |
|---|---|---|---|---|
| **A** | Back | 21⁄32 × 18 × 26½" | 1 | Pine panel |
| **B** | Side | 21⁄32 × 3⅝ × 13" | 2 | Pine panel |
| **C** | Shelf | 21⁄32 × 2¾ × 26½" | 1 | Pine panel |

**Materials:** #6 × 1½" wood screws, 4d finish nails, 2 steel keyhole hanger plates.

**Note:** Measurements reflect the actual thickness of dimensional lumber.

*Lay out the pattern on the back piece by tracing half the pattern on one side of the centerline, then flipping the pattern and tracing the other side.*

*Clamp the back to your worksurface, and cut the pattern, using a jig saw.*

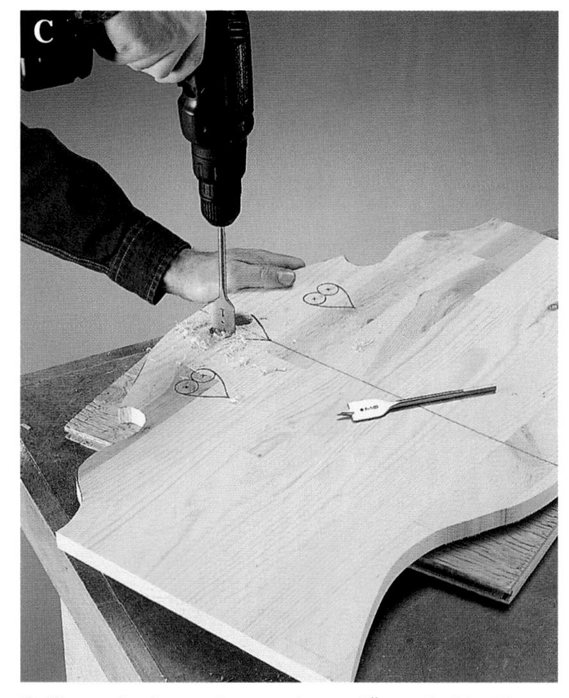

*Drill out the larger heart using a 1" spade bit. Use a ⅞" spade bit to create the smaller hearts.*

## Directions: Plate & Spoon Rack

CUT THE BACK. Laying out the back requires the use of a pattern (see *Diagram*). Our pattern includes half of the back design, and will be flipped to mark two symmetrical halves. Cut the back blank (A) to overall size. Transfer the decorative profile onto cardboard or heavy paper (see *Diagram*). Locate the back piece centerline and work both ways, tracing the pattern on one side of the centerline, then flipping the pattern over at the centerline and tracing the other symmetrical side **(photo A).** Cut the edge shapes with a jig saw, taking care that the ends, which will be attached to the sides, are finished at 12½" **(photo B).**

CUT THE HEART DETAILS. The back has two different-sized heart cutouts (see *Diagram*). Trace these heart cutouts onto the back piece. Place a piece of scrap under the workpiece to protect your worksurface, and start by drilling two horizontally adjacent holes, using a 1" spade bit for the larger heart and a ⅞" bit for the smaller hearts **(photo C)**. Complete the cutouts with a jig saw.

CUT THE SIDES. Cut the sides (B) by transferring the side profile to cardboard (see *Diagram*) and then tracing onto pine. Cut one side with a jig saw, then use the completed side as a pattern to mark and cut the other side.

CUT THE SHELF. The shelf has a groove to hold the plates, and notches to hold the spoons.

*Cut the plate groove into the shelf, using bar clamps and a straightedge to hold the shelf in place and guide the router.*

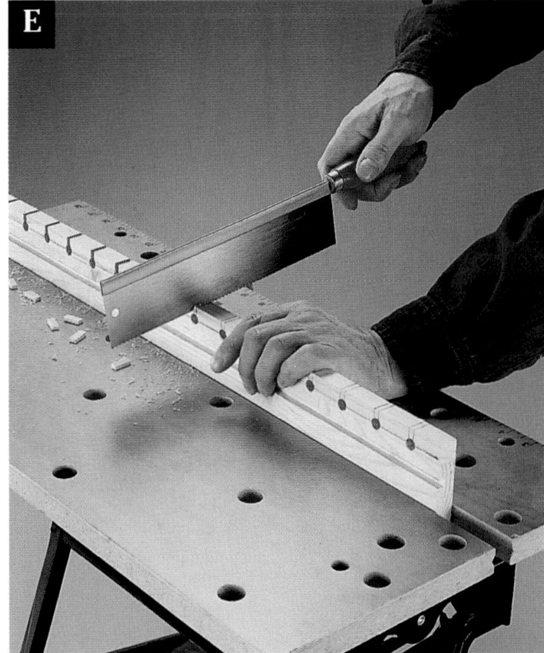

*Cut a 3/16"-wide notch along the front edge for each of the spoon rack holes, using a backsaw.*

First cut the shelf (C) to size. Then cut the groove into the shelf, using a simple routing jig to help make a uniform, straight cut. To make the jig, clamp two pieces of 3/4" scrap wood against the edges of the shelf. Securely clamp a straightedge over the scrap wood to guide the router base and cut the groove at the desired location (see *Diagram*) with a 1/4" straight router bit set 1/4" deep **(photo D).** Next, measure and mark the 17 spoon hole centerpoints (see *Diagram*). Drill the holes, using a 3/8" bit. Complete the spoon hole cutouts, using a backsaw to cut straight into the holes, leaving a 3/16" slot **(photo E).** Sand all cut edges before assembling.

ASSEMBLE THE RACK. Drill pilot holes into the sides and attach to the back using glue and 4d finish nails. Position the shelf, then drive 4d finish nails through the sides into the shelf **(photo F).** Next, drill countersunk pilot holes and drive wood screws through the back into the shelf.

### TIP

*Pine wood actually sands better with a vibrating sander than a belt sander because pitch typically builds up faster in the belt and causes burn marks or unevenness.*

APPLY FINISHING TOUCHES. Sand smooth and apply a finish. We recommend amber shellac, which provides a hard, cleanable surface and gives the wood a very warm, rich and even color.

HANG THE RACK. Drill or chisel out space so the keyhole hanger plates are flush with the back surface of the rack. Drive screws through the hanger plates into the back of the rack.

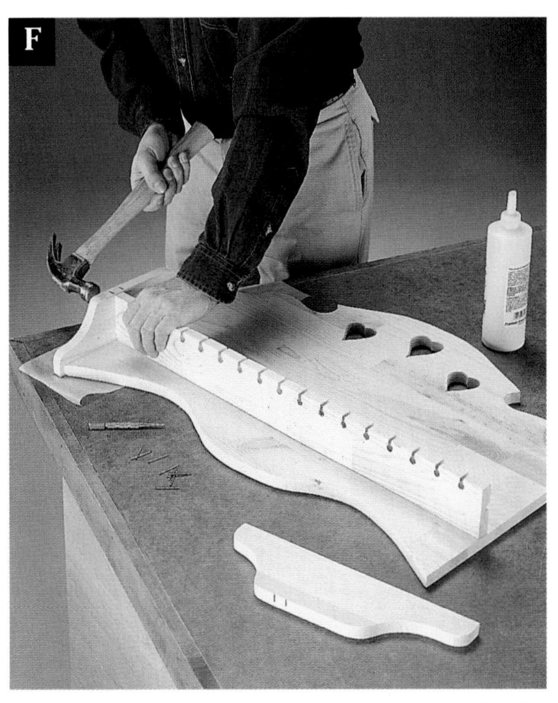

*Attach the sides to the back with glue and 4d finish nails, then attach the shelf to the sides with finish nails and to the back with wood screws.*

# Vertical Display Cabinet

*Sturdy but airy, this open display unit combines the warmth of oak with the glimmer of glass.*

PROJECT POWER TOOLS

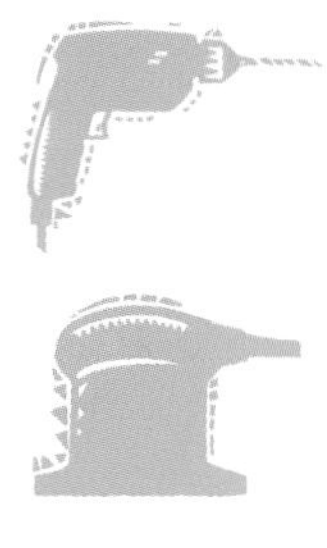

CONSTRUCTION MATERIALS

| Quantity | Lumber |
|---|---|
| 1 | 1 × 1" × 8' oak |
| 12 | 1 × 2" × 8' oak |
| 2 | 1 × 4" × 8' oak |
| 1 | 1 × 4" × 4' oak |
| 1 | 1 × 6" × 8' oak |
| 1 | ¾" × 4 × 4' oak plywood |
| 1 | 3½" × 8' oak crown |
| 1 | 8' oak base shoe |
| 3 | ⅜ × 16¾ × 39¼" tempered glass |

Compare the design and price of our vertical display cabinet to those available in stores, and you'll be impressed. Though the materials are not expensive, this project is by no means cheap in construction. The slender stiles are made from oak hardwood for strength. The oak back braces perform double duty by adding an unusual visual element as well as reinforcing the cabinet. Transparent glass shelves give the cabinet a modern appeal, while oak crown and base trim pieces lend classic elegance.

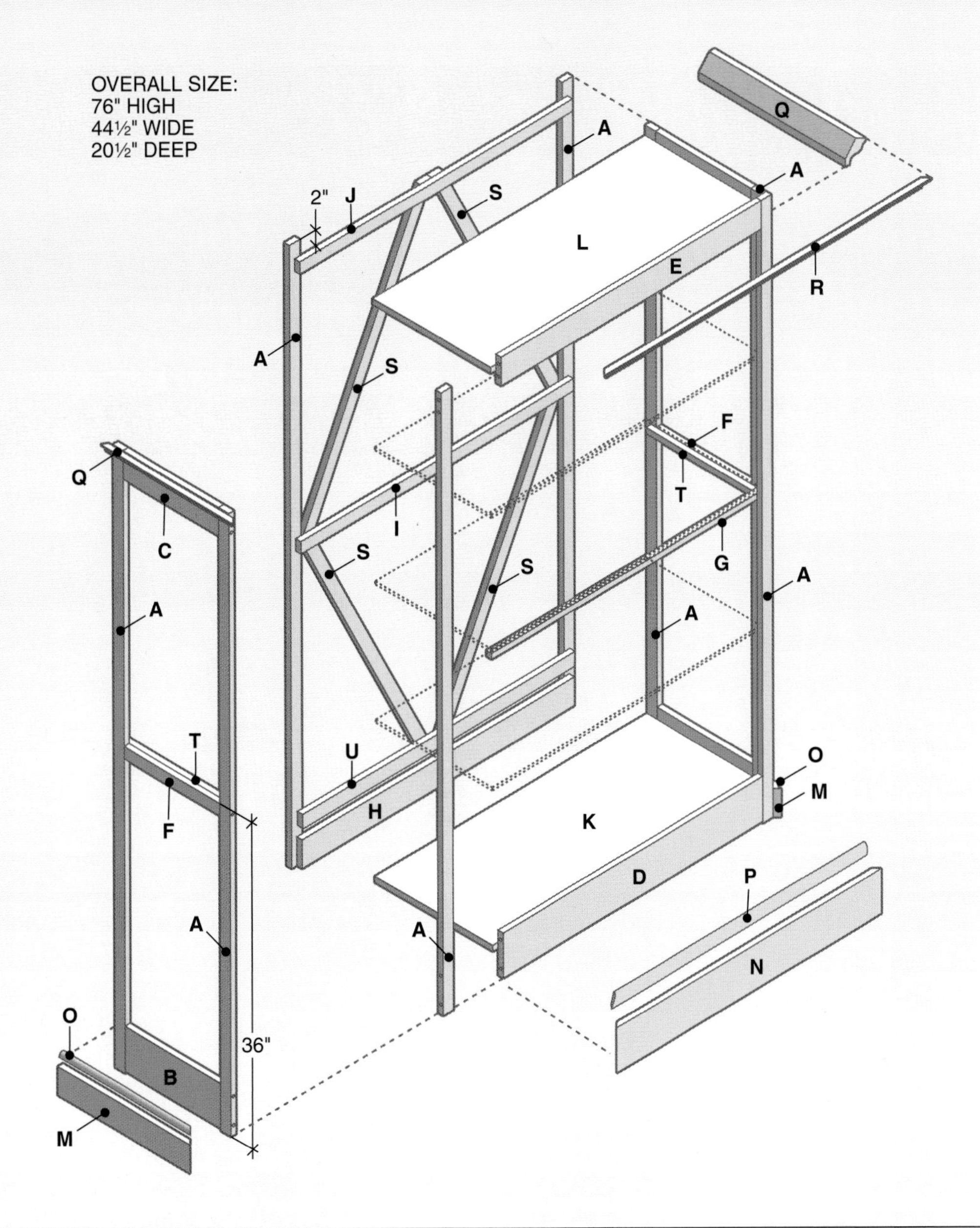


| Key | Part | Dimension | Pcs. | Material |
|---|---|---|---|---|
| A | Stile | ¾ × 1½ × 76" | 8 | Oak |
| B | Short bottom rail | ¾ × 5½ × 14" | 2 | Oak |
| C | Short top rail | ¾ × 3½ × 14" | 2 | Oak |
| D | Long bottom rail | ¾ × 5½ × 38" | 1 | Oak |
| E | Long top rail | ¾ × 3½ × 38" | 1 | Oak |
| F | Center rail | ¾ × 1½ × 14" | 2 | Oak |
| G | Front shelf rail | ¾ × ¾ × 39½" | 1 | Oak |
| H | Back bottom cleat | ¾ × 3½ × 39½" | 1 | Oak |
| I | Back shelf rail | ¾ × 1½ × 39½" | 1 | Oak |
| J | Back top cleat | ¾ × 1½ × 39½" | 1 | Oak |
| K | Bottom | ¾ × 17 × 39½" | 1 | Oak ply. |

**Cutting List**

| Key | Part | Dimension | Pcs. | Material |
|---|---|---|---|---|
| L | Top | ¾ × 17 × 39½" | 1 | Oak ply. |
| M | Side base | ¾ × 3½ × 19¼" | 2 | Oak |
| N | Front base | ¾ × 3½ × 42½" | 1 | Oak |
| O | Side base cap | ⅜ × ¾ × 19¼" | 2 | Oak base shoe |
| P | Front base cap | ⅜ × ¾ × 42½" | 1 | Oak base shoe |
| Q | Side crown mldg. | ¾ × 3½ × 21" | 2 | Oak crown |
| R | Front crown mldg. | ¾ × 3½ × 44½" | 1 | Oak crown |
| S | Back brace | ¾ × 1½ × 39¾" | 4 | Oak |
| T | Center shelf cleat | ¾ × ¾ × 15⅝" | 2 | Oak |
| U | Bottom stretcher | ¾ × 1½ × 39½" | 1 | Oak |
| | | | | |

**Materials:** Wood glue, #6 wood screws (1¼", 1½"), 4d finish nails, 1" brads, ⅜" × 4' oak doweling (2), ⅜" oak plugs, plastic pin-style shelf supports (12), self-adhesive shelf cushions, finishing materials.

**Note:** Measurements reflect the actual thickness of dimensional lumber.

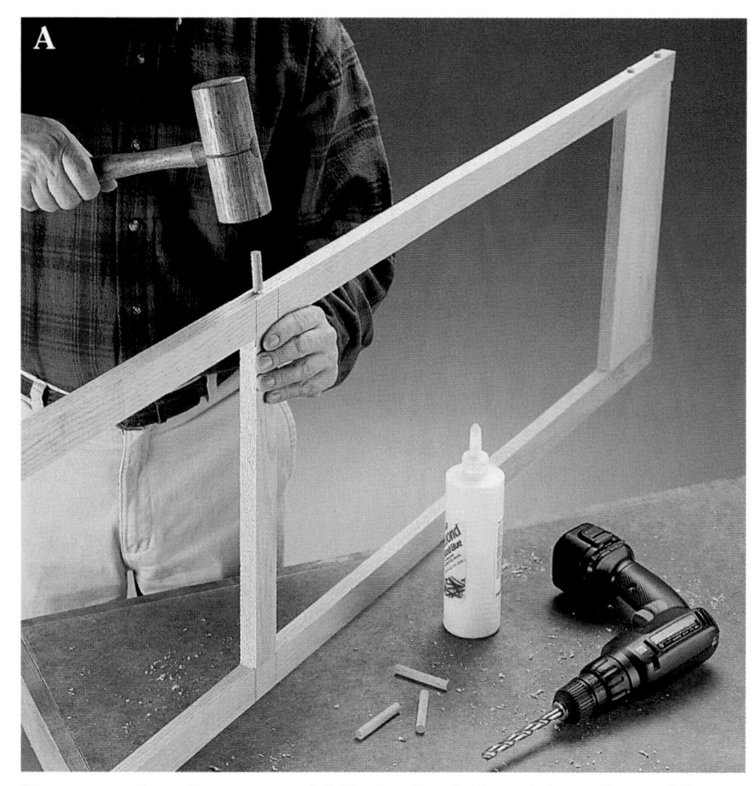
Use a wood mallet to pound ⅜" glued oak dowels into the end frame.

Measure the diagonals to check for square. If the measurements are equal, the frame is square.

## Directions: Vertical Display Cabinet

MAKE THE END FRAMES. Cut the stiles (A), short bottom rails (B), short top rails (C) and center rails (F) to size, and sand smooth. On each stile, make a reference mark 36" down from the top edge. Position a top short rail between two stiles with the top edges flush. Position a bottom rail ¼" up from the bottom of the stiles, and place the center rail so the top edge lies on the 36" mark.

Cut 20 lengths of ⅜" dowling, 2½" long. Score the edges of the dowels to make them easier to insert (see *Tip*, page 42). Drill ⅜ × 2½"-deep holes through the stiles into the rails (see *Diagram*). Apply glue to the joints and drive dowels into the holes **(photo A).** Check to make sure the end frames are square, then clamp the pieces together until the glue dries.

MAKE THE FRONT FRAME. Cut the long bottom rail (D) and long top rail (E) to size, and sand smooth. Position the rails between the two stiles (the top rail should be flush, the bottom rail should be ¼" up from the ends of the stiles). Drill holes and attach the rails and stiles with glued ⅜" dowels.

Cut the front shelf rail (G) to size, and sand smooth. Lay the front frame facedown, then position the front shelf rail so the top edge is flush with the 36" reference lines on the stiles and the ends are set back ¾" from the edges of the stiles. Drill countersunk pilot holes and fasten the shelf rail with glue and 1¼" screws.

MAKE THE BACK FRAME. Cut the back bottom cleat (H), the back shelf rail (I) and the back top cleat (J) to size, and sand smooth. Lay two stiles on your worksurface, and position the bottom cleat over the stiles so the bottom edges are flush and the ends of the cleat are set in ¾" from the edges of the stiles. Drill countersunk pilot holes and attach the cleat with glue and 1¼" screws.

Position the back top cleat over the stiles so the top edge is 2" down from the top of the stiles and the ends are set in ¾" from the edges. Attach the top cleat with glue and countersunk screws. Measure the diagonals of the frame to check for square **(photo B).**

ASSEMBLE THE CABINET. Position the end frames upright on their back edges, then lay the front frame over them (the stiles on the end frames should fit tightly against the inset cleats on the front frame). Check for square by measuring diagonals, then drill counterbored pilot holes and join the front

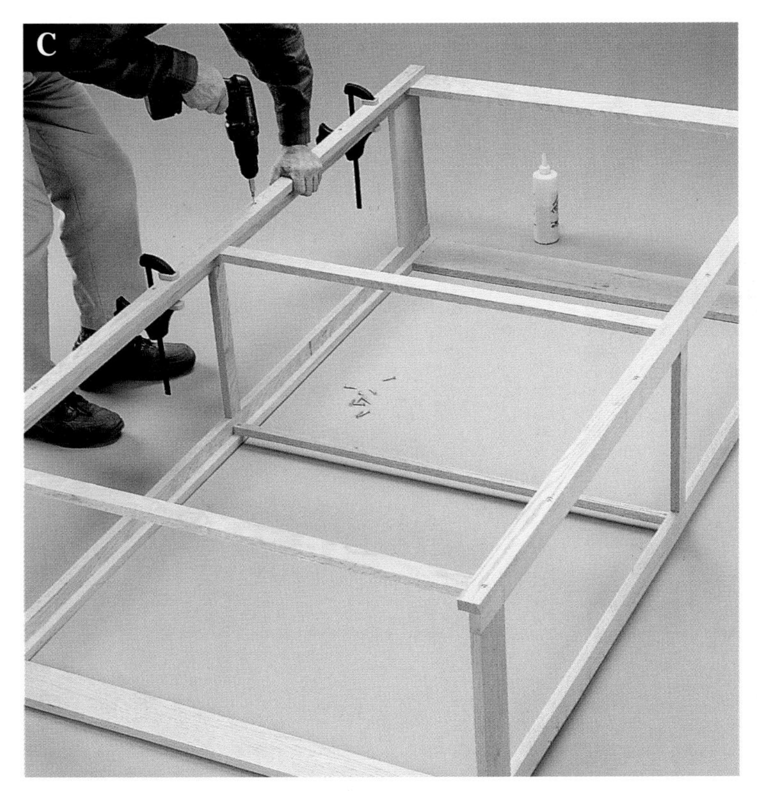


*Position the back over the end frames at the ¾" space, and join with glue and screws.*

*Secure the back braces at the centerpoints to create a diamond.*

frame to the end frames with glue and 1½" screws.

Turn the assembly over so the front frame faces down, then position and attach the back frame with glue and counterbored 1½" screws **(photo C).**

ATTACH THE BOTTOM, TOP AND STRETCHER. Cut the bottom (K) and top (L) to size, and sand the edges smooth. Position the bottom piece in the cabinet so its bottom surface is set 3½" up from the lower ends of the stiles (at the rear, the bottom piece will rest on the back cleat). Drill countersunk pilot holes through the bottom rails and into the bottom, then attach with glue and 1½" screws.

Fasten the top in similar fashion. First, position the top piece in the cabinet so the top surface is 1¼" below the top ends of the stiles (at the back, the top piece will rest on the back cleat). Drill countersunk pilot holes and attach the top with glue and 1½" screws.

Cut the bottom stretcher (U) to size, and sand smooth. Position it over the rear edge of the bottom piece, then attach it to the stiles with glue and countersunk 1¼" screws.

ATTACH THE BACK BRACES. The easiest way to cut the angled back braces is to position the 1½" stock for the first brace across the back of the cabinet, mark and cut the proper angles, then use this piece as a template for cutting the remaining braces.

Begin by cutting blanks for the back braces (S) to length. Measuring horizontally, mark the centerpoints of the back stretcher and back top cleat. On each end of the back shelf rail, mark vertical centerpoints ¾" inch from the bottom edge.

Position a back brace diagonally from the back top cleat to the back shelf rail, with the ends touching the reference marks. Use a T-bevel to mark horizontal cutting lines across the ends of the brace, parallel to the edges of the top cleat and shelf cleat. Cut the angles on the ends of the brace, then use this back brace as a template to cut the other three.

Drill countersunk pilot holes, and attach the back braces with glue and 1¼" screws **(photo D).**

ATTACH THE TRIM. Cut all base pieces (M, N), cap pieces (O, P) and crowns (Q, R) to length, mitering the appropriate ends at 45° **(photo E).**

Position the base pieces around the front and sides of the cabinet, flush with the bottom ends of the stiles, then drill pilot holes and attach with 4d finish nails. Position the caps on the base pieces, drill

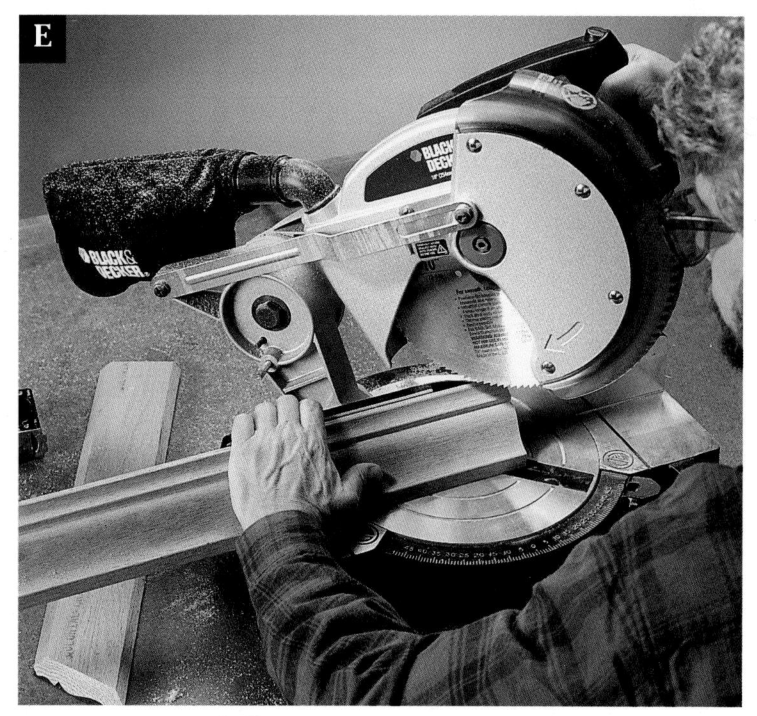


*Miter the crown molding by positioning the molding upside down in the miter box. Cut the moldings about ½" too long, then test-fit the pieces and trim as needed for a tight fit at the mitered corners.*

*Attach the center shelf cleats to the center rails with glue, and clamp the pieces together.*

pilot holes and secure with glue and brads.

Position the crown moldings so the bottom edges are 1½" from the bottom of the top rails, drill pilot holes and attach with 4d finish nails. Lock-nail the mitered joints.

ADD THE SHELF SUPPORTS. Cut the center shelf cleats (T) to size, and sand smooth. Apply glue, then position the cleats against the center rails so the top edges are flush, and clamp in place until the glue dries **(photo F).**

To position the shelf support holes, cut an 8 × 20" pegboard template. Outline a horizontal row of holes 16" up from the bottom. Resting the template on the top edge of the bottom rails and bottom stretcher, drill holes in the back braces and stiles **(photo G).** Drill holes for the upper shelf in similar fashion, resting the template on the center rails and back shelf rail.

APPLY FINISHING TOUCHES. Insert ⅜" glued oak plugs into counterbores, and apply putty to visible nail holes. Finish-sand, then apply the finish of your choice (we used walnut stain). Attach self-adhesive plastic cushions to the shelf supports, then install the glass shelves.

*Use a pegboard template to align the shelf support holes in the stiles and back braces.*